CHEMICAL
FOOD
SAFETY

CHEMICAL FOOD SAFETY

A Scientist's Perspective

Jim
Riviere,
D.V.M., Ph.D.

Iowa State Press
A Blackwell Publishing Company

JIM RIVIERE, D.V.M., PH.D., is the Burroughs Wellcome Fund Distinguished Professor of Pharmacology and Director, Center for Chemical Toxicology Research and Pharmacokinetics, FAHRM Department, College of Veterinary Medicine, North Carolina State University, Raleigh, N.C. He is a co-founder and co-director of the USDA Food Animal Residue Avoidance Databank (FARAD). Dr. Riviere presently serves on the Food and Drug Administration's Science Board and on a National Academy of Science panel reviewing the scientific criteria for safe food. His current research interests relate to risk assessment of chemical mixtures, the pharmacokinetics of tissue residues and absorption for drugs and chemicals across skin.

© 2002 Iowa State Press
A Blackwell Publishing Company
All rights reserved

Iowa State Press
2121 State Avenue, Ames, Iowa 50014

Orders: 1-800-862-6657
Office: 1-515-292-0140
Fax: 1-515-292-3348
Web site: www.iowastatepress.com

First edition, 2002

Library of Congress Cataloging-in-Publication Data

Riviere, J. Edmond (Jim Edmond)
    Chemical food safety : a scientist's perspective / Jim Riviere.—1st
ed.
    p. cm.
Includes bibliographical references and index.
ISBN 0-8138-0254-7 (alk. paper)
    1. Food contamination. 2. Food additives—Toxicology. 3. Health risk
    assessment. I. Title.
TX531 .R45 2002
363.19′2—dc21

                                                        2002003498

The last digit is the print number: 9 8 7 6 5 4 3 2 1

To Nancy,
Whose love makes it worth living!

Contents

Acknowledgements, ix
Introduction, xi

1 Probability—The Language of Science and Change, 3
2 Dose Makes the Difference, 19
3 The Pesticide Threat, 37
4 Veggie and Fruit Consumption in the Age
      of the Centenarians, 55
5 Chemicals Present in Natural Foods, 69
6 Risk and Regulations, 83
7 Some Real Issues in Chemical Toxicology, 91
8 Milk Is Good for You, 111
9 Biotechnology and Genetically Modified Foods:
      "The Invasion of the Killer Corn" or "New Veggies
      to the Rescue", 121
10 Food Security and the World of Bioterrorism, 135
11 The Future, 145

A Toxicology Primer, 159
B Selected Readings and Notes, 179

Index, 199

Acknowledgements

This work is a revised and expanded version of my original book on this subject: *Why Our Food is Safer Through Science: Fallacies of the Chemical Threat*, published in 1997. The lion's share of thanks for both works must go to Dr. Nancy Ann Monteiro (my spouse, soul mate, and fellow toxicologist) for truly helping me write these books and otherwise putting up with me while I did it. Her insights, honest criticisms, and suggestions for topics and approaches remain indispensable. I thank our children, Christopher, Brian, and Jessica, for being there and tolerating lost time together. I especially thank our youngest, Jessica, for adopting a diet that forced me to look into this issue.

There are numerous professional associates who have contributed to these works. I thank Dr. Arthur Craigmill, a close and true friend who continues to "feed me" with topics that strengthen this book. I am indebted to the original reviewers of this work, as much of it continues to exist in the present version. I thank colleagues and students at North Carolina State University for untold discussions and perspectives and for tolerating me while I was immersed in this book. I am grateful to my administrators for giving me the time to write. I thank Mr. David Rosenbaum of Iowa State Press for deciding to publish this updated work. Finally, I must again thank the editors of *Newsweek* for publishing the original essay that got me into this frame of mind to begin with.

Introduction

This is a book about the potential danger of pesticide residues in our food and how science can be applied to make these risk assessments. It is on the practical application of science to chemical food safety. I originally tackled this topic in 1997, in the predecessor to this book: *Why Our Food is Safer Through Science: Fallacies of the Chemical Threat.* Writing this book opened my eyes to numerous fallacies of the public perception of risk assessment and of their intolerance to certain types of risks, no matter how remote. As this book was being revised, the tragic events of September 11 unfolded, along with the new threat of bioterrorism as anthrax was detected in letters to news organizations and government officials. This author was involved in debriefing some government officials since my research has been focused for many years on military chemicals. As will become evident throughout this book, my other professional focus is on chemical food safety. I never expected that these two endeavors would merge into a sole concern. In fact, in the first incarnation of this book, I made the argument that these chemicals, used in military warfare, are truly "toxic" and would deserve attention if one were ever exposed to them. I had considered, but never believed, that they could actually be a potential for their use against American citizens. I have added a chapter on this topic as it further illustrates how "real risks" should be differentiated from the many "phantom risks" presented in this book. The phobia against chemicals and infectious diseases, rampant in the American populace, unfortunately greatly plays into the terrorist's hands because panic magnifies small incidents of exposure to epidemic proportions.

I still cannot believe that many of the points on chemophobia and risk assessment present in the 1997 book still need to be addressed! A sampling of the following headlines attests to the media confusion that is present, regarding whether the health of modern society is improving or deteriorating.

Improving?	or	Deteriorating?
Odds of Old Age Are Better than Ever		Herbicides on Crops Pose Risk
U.S. Posts Record Gain in Life Expectancy		Chemicals Tinker with Sexuality
Kids Who Shun Veggies Risk Ill Health Later		Pesticides Found in Baby Foods
Americans Eat Better, Live Longer		Biotech Corn Found in Taco Shells
EPA Reports Little Risk to Butterfly from Biotech Corn		Monarch Butterfly Doomed from Biotech Crop

What is going on here? Should we be satisfied and continue current practices or be concerned and demand change through political or legal action?

These stories report isolated studies that if closely compared suggest that diametrically opposed events are occurring in society. Some stories suggest we are living longer because of modern healthcare, technology, and increased consumption of a healthy food supply. Others express serious concern for our food supply and environment and indicate something might need to be corrected before disastrous outcomes befall us all.

What is the truth? I strongly believe that the evidence is firmly in the court of the positive stories. I especially believe the data that prove eating a diet rich in vegetables and fruits is one of the healthiest behaviors to adopt if one wants to minimize the risk of many diseases.

These conflicting headlines strike at the heart of my professional and personal life. I am a practicing scientist, veterinarian, and professor working in a public land-grant university. My field of expertise can be broadly defined as toxicology and pharmacology. I am "Board Certified" in Toxicology by the Academy of Toxicological Sciences. To further clear the air, I am not financially supported by any chemical manufacturing company. The bulk of my research is funded from federal research grants. I have worked and published on chemical factors that may impact the Gulf War illness and on the absorption of pesticides and environmental contaminants applied in complex chemical mixtures. I have been involved for twenty years with an outreach/ extension program aimed at avoiding drug and chemical residues in milk and meat.

I have always felt comfortable in the knowledge that the basic premises underlying my discipline are widely accepted and generally judged valid by my colleagues and other experts working and teaching in this area. These fundamentals are sound and have not really changed in the last three decades.

When applied to medicine, they have been responsible for development of most of the "wonder drugs" that have eliminated many diseases that have scourged and ravaged humanity for millennia. Yet even in these incidences, some serious side effects have occurred, and formerly irradicated diseases are making their presence felt.

I am also the father of three children and, like all parents, have their welfare in mind. Thus I am concerned with their diet and its possible effects on their long-term health. I am both a generator and consumer of food safety data. In fact, I originally started down this trail when the "father" in me watched a television news magazine on food safety and the dangers of pesticides in our produce. The program also highlighted the apparent dangers of biotechnology. This program was far from being unique on this topic, and similar ones continue to be aired even to this day. However, the "scientist" in me revolted when almost everything presented was either taken out of context, grossly misinterpreted, or was blatantly wrong.

This revelation prompted me to look further into the popular writings on this issue. I became alarmed at the antitechnology slant of much of the media coverage concerning food safety. As a result, I wrote an essay that was published in *Newsweek*. The comments I received convinced me that a book was needed to reinforce to the public that our food is healthier and safer than ever before in history. This article and the television program that prompted it occurred in early 1994. The headlines quoted above appeared in 1995 and some in 2000! Obviously the problem of misinformation is still present, and there is no indication that it will abate. Two books published in 2001, *It Ain't Necessarily So* by David Murray, Joel Schwartz and S. Robert Lichter as well as *Damned Lies and Statistics* by Joel Best, continue to illustrate that many of the problems presented in this book are relevant and must be addressed as we encounter new challenges.

I remain concerned that "chemophobia" is rampant and the scientific infrastructure that resulted in our historically unprecedented, disease-free society may be inadvertently restructured to protect a few individuals from a theoretical and unrealistic risk or even fear of getting cancer. Also, I am concerned because our government has funded what should be definitive studies to allay these fears, yet these studies and their findings are ignored!

The most troublesome discovery from my venture into the popular writings of toxicology and risk assessment is an underlying assumption that permeates these books and articles, that the trained experts are often wrong and have vested interests in maintaining the *status quo*. Instead of supporting positions with data and scientifically controlled studies, the current strategy of many is to attack by innuendo and anecdotes, and to imply guilt by association with past generations of unknown but apparently mistaken scientists. Past predictions are never checked against reality, much as the forecasts of the former gurus of Wall Street are not held accountable when their rosy predictions missed the stock market tumble of the last year.

The overwhelming positive evidence of the increase in health and longevity of today's citizens is ignored. Instead, speculative dangers and theoretical risks are discussed *ad nauseum*. The major problem with this distraction is that valuable time and resources are diverted from *real* medical issues that deserve attention.

In an ideal world, all scientists and so-called "expert witnesses" would subscribe to the "Objective Scientists Model" from the American College of Forensic Examiners, which states that:

> Experts should conduct their examinations and consultations and render objective opinions regardless of who is paying their fee. Experts are only concerned with establishing the truth and are not advocates. Forensic Examiners, whether they work for the government or in the private sector, should be free from any pressure in making their opinion.

In such an ideal world, experts would state the truth when they know it and refrain from making opinions when there is insufficient data. This, of course, assumes that experts have the necessary credentials to even comment. As this book will show, we obviously do not live in an ideal world!

Why am I concerned enough about these issues to write this book? Because I do not want to live in a society where science is misused and my health is endangered because of chemophobia and an irrational fear of the unknown. While writing the original version of this book, Philip Howard's *The Death of Common Sense: How Law is Suffocating America* was published. After reading it, I realized that as a practicing scientist, I have an obligation to make sure that the data I generate are not only scientifically correct but also not misinterpreted or misused to the detriment of the public. By being silent, one shares responsibility for the outcome. The two new books by Murray and Best, mentioned earlier, give more examples of how our public views risk assessment and, furthermore, how it is a scientist's obligation to help educate the public on the use of scientific information in the arena of public policy.

There are numerous examples of how common sense is dead. Philip Howard quotes some interesting cases relating to chemical regulations gone awry that remain pertinent today. The most amusing, to his readers (but not the individuals affected), is the requirement that sand in a brick factory be treated as a hazardous chemical since it contains silica which is classified by our regulators as a poison. I am writing this book from the sandy Outer Banks of North Carolina where I am surrounded by this toxic substance! Should I avoid going to the beach? How did and do these regulations continue to get derailed? Whose job is it to put them back on track?

Similar absurdities abound everywhere. In my own university, because of concern for insurance liability, our photocopy machine was removed from an open hallway due to fire regulations and the fear of combustion! It is now located in a closed conference room and is inaccessible when meetings are in progress. Similarly, paper notices are prohibited from being posted in corridors for fear of fire. I guess I should feel comforted by the fact that the same prohibition was mentioned in Howard's book in a school system where kindergarten classes were prohibited from posting their artwork. Do such regulations protect the safety of our children? A similar position is the view that pesticide residues in food, at the levels existing today, are actually harmful to human health. Some would argue that we should be worried about dying of cancer if we eat an apple with trace levels of a pesticide. This book will argue that this is a theoretical and nonsignificant risk. There are far more serious health problems that merit our attention and deserve our resources to irradicate.

I do credit myself with one small success in this regard. I was the member of my university's Hazardous Material Committee some years ago when we were trying to establish transportation guidelines for moving "poisons" across campus. The intent of the regulations was to protect staff and students when "truly toxic" chemicals were being transported. We wanted to ensure that a student's "backpack" did not become a "hazardous material transport container." However, we soon discovered that just because a compound appeared on a list of hazardous materials, it does not mean that special precautions need to be established for its handling. The most poignant example was the regulation, which if followed to the "letter of the law" would have required the university to employ a state-certified pesticide expert to transport a can of the pesticide Raid® from the local supermarket to my laboratory. Let's get real!

These overinterpretations cause serious problems because they dilute the impact of regulations designed to protect people from real hazards. If one is made to follow too many regulations for handling chemicals that present minimal risks, these precautions will become routine. Then when a truly hazardous chemical is handled, there is no way to heighten awareness and have the person exercise real caution. As will become obvious, there are hazards that need our attention, and these misdirected concerns channel our attention in the wrong direction. Everyone becomes confused. I have conducted research with many of the pesticides that will be discussed in this book. I have served on various advisory panels including some for the Environmental Protection Agency (EPA) and the Food and Drug Administration (FDA). Thus, I have some working knowledge of how these regulations come into being and are put into practice. I also unfortunately see how they can be misused.

In our laboratory, we routinely deal with "really" toxic chemicals including a number of very potent anticancer drugs, pesticides, and high doses of a whole slew of other poisons including chemical warfare agents. I want to make sure that when a student or colleague works with one of these "truly hazardous" chemicals, that this real risk is not trivialized and additional precautions are followed that are not required when picking up a can of Raid®! This is the true "adverse effect" resulting from this overreaction to trivial chemical risks.

The letters to the editor that resulted from my original *Newsweek* essay and comments after the publication of the original version of this book only further support Philip Howard's contention that "common sense is dead." In one phrase of the essay, I implied that pesticides are harmless. The point of the article (and this book) was that they are essentially harmless at the low levels found in the diet. Responders took this phrase out of context and attacked! They suggested that what I wrote was biased because my laboratory must be "heavily supported" by the chemical industry. (However much I would wish, this is not true!) In any event, this should not be relevant because I take the above "Objective Scientist Model" seriously. Second, all research is supported by someone and thus biases due to the supporting agency/advocacy group or the individual scientist may always be present.

Another writer suggested that what I wrote could never be trusted because I am from the state of North Carolina and thus as a "Tarheel" must have a track record of lying in support of the safety of tobacco! This is a ludicrous argument even if I were a native Tarheel. My North Carolina friends and colleagues working in the Research Triangle Park at the National Institute of Environmental Health Science and the Environmental Protection Agency would surely be insulted if they knew that their state of residence were suddenly a factor affecting their scientific integrity! Such tactics of destroying the messenger are also highlighted in Murray's book *It Ain't Necessarily So.* Such attacks, which have nothing to do with the data presented, convinced me that I had an obligation to speak up and say, "Let's get real!"

This book deals with the potential health consequences of pesticides, food additives, and drugs. All of these are chemicals. The sciences of toxicology, nutrition, and pharmacology revolve around the study of how such chemicals interact with our bodies. As will be demonstrated, the only factor which often puts a specific chemical under the umbrella of one of these disciplines is the source and dose of the chemical or the intent of its action. The science is the same in all three. This is often forgotten!

If scientists such as myself who know better keep quiet, then someone will legislate or litigate the pesticides off of my fruit and vegetables, making this essential food stuff too expensive or unattractive to eat or ship. They will remove chlorine from my drinking water for fear of a theoretical risk of cancer while increasing the real risk of waterborne diseases such as cholera. They will outlaw biotechnology, which is one of our best strategies for actually reducing the use of synthetic chemicals on foods and generating more user-friendly drugs to cure diseases that presently devastate us (AIDS, cancer) and those which have yet to strike (e.g., the emerging nightmares such as Lassa fever and the Ebola virus and hantaviruses). By banning the use of pesticides and plastics in agriculture, our efficiency of food production will decrease which may result in the need to devote increased acreage to farming, thus further robbing our indigenous wild animal populations of valuable habitat. Our environment will suffer.

I have attempted to redirect the present book to those whose career may be in food safety, risk assessment, or toxicology because they are often unaware of how the science of these disciplines may be misinterpreted or misdirected. These issues are not covered in academic and scholarly textbooks. Conclusions published in the scientific literature may be misdirected to support positions that the data actually counter. This may often occur simply because the normal statement of scientific uncertainty (required to be stated when any conclusion is made in a scientific publication) is hijacked as the main findings of the study. These pitfalls must be brought to the attention of working scientists so conclusions from research studies can be crafted to avoid this pitfall.

The central thesis of this book is as follows: The evidence is clear that we have created a society that is healthier and lives longer than any other society in history. Why tinker with success? We have identified problems in the past and have corrected them. The past must stay in the past! One cannot continuously raise the specter of DDT when talking about today's generation of pesticides. They are not related. If one does use this tactic, then one must live with the consequences of that action. When DDT usage was reduced in some developing countries because of fear of toxicity, actual deaths from malaria dramatically increased. DDT continues to be extensively used around the world with beneficial results.

This book is not about the effect of pesticide spills on ecology and our environment. These are separate issues! This book is about assessing the risk to human health created by low-level pesticide exposure. There are serious problems in the world that need our undivided attention. The theoretical risk of one in one million people dying from cancer due to eating an apple is not one of them, and continued attention to these "nonproblems" will hurt us all. In fact, using readily available data, one could calculate a much higher risk from *not* eating the apple since beneficial nutrients would be avoided! This is why I wrote this book.

CHEMICAL FOOD SAFETY

Probability—The Language of Science and Change

There are lies, damned lies, and statistics.

(Benjamin Disraeli)

This book is written to refute the current public misconception that modern science and technology have created an unsafe food supply through widespread contamination with pesticides or unbridled biotechnology. Although mistakes may have been made in the past, most have been corrected long ago and cannot be continually dragged up and used to support arguments concerning the safety of today's products. Eating fruits and vegetables is actually good for you, even if a very small percentage have low levels of pesticide residues present. The scarce resources available for research today should be directed to other issues of public health and not toward problems that do not exist. Most importantly, advocating the elimination of chemical residues in food while simultaneously attacking modern advances in biotechnology aimed at reducing this chemical usage are diametrically opposed causes. Both should not be championed by the same individual because the best hope for reducing the use of organic chemicals is through bioengineering!

The student entering the field of food science is often not aware of these societal biases against chemicals and biotechnology. The excitement of the underlying scientific breakthroughs in genomics and other fields is naturally applied to developing new food types that fully use the safer technologies that form the basis of these disciplines. Thus genes are manipulated in plants to produce enzymes that prevent attack by pests without the need to apply potentially toxic chemicals. When these truly safer foods are brought to market, students are amazed to discover the animosity directed against them for using technology to manipulate food. It is this thought process that must be confronted.

The primary reason these issues attract such widespread attention is that modern Western society, especially in the United States, believes that science and technology are in control and disease should not occur. If anything goes wrong, someone must be to blame. There is no doubt we are living longer than ever before—a fact often conveniently ignored by many as it obviously does not support a doomsday scenario. Just ask any member of our burgeoning population of centenarians. As we successfully combat and treat the diseases that once killed our ancestors and grandparents at an earlier age, we will obviously live longer, and new disease syndromes will be encountered as we age. They will become our new cause of death. Yet since we believe that this shouldn't happen, we must blame someone, and modern technology is often the focus of our well-intentioned but misplaced angst over our impending death. The reader must keep in mind that we all will die from something. It is the inevitable, if not obvious, final truth. The topic of this book is whether eating pesticides on treated fruits or vegetables has an effect on *when* and *how* this ultimate event will occur.

A major source of confusion in the media relates to how science tries to separate unavoidable death due to natural causes from avoidable death due to modern drugs or chemicals. This is the field of risk assessment and is a crucial function of our regulatory agencies. Confusion and unnecessary anxiety arise from the complexity of the underlying science, misinterpretation of the meaning of scientific results, and a lack of understanding of the basic principles of probability, statistics, and causal inference (assigning cause and effect to an event).

The principles to be discussed comprise the cornerstone of modern science, and their continuous application has resulted in many benefits of modern medicine that we all enjoy. Application of statistics to science and medicine has most likely contributed to the increased human lifespan observed in recent decades. However, more recently, it has also been responsible for much of the concern we have over adverse effects of modern chemicals and pesticides. Since all of the data to answer these issues are expressed as probabilities, using the language of statistics, what does using statistics really mean? How should results be interpreted?

Scientists are well aware that chance alone often may determine the fate of an experiment. Most members of the public also know how chance operates in their daily lives but fail to realize that the same principles apply to most scientific studies and form the core of the risk-assessment process. Modern science and medicine are based on the cornerstone of reproducibility because there is so much natural variation in the world. One must be assured that the results of a scientific study are due to the effects being tested and not simply a quirk of chance. The results should also apply to a majority of the population. *Statistics is essentially the science of quantifying chance and determining if the drug or chemical tested in an experiment changes the outcome in a manner that is not due to chance alone.* This is causality as operationally defined in statistics, risk assessment, and many epidemiological studies. It does not assign a mechanism to this effect.

Two problems arise in evaluating chance in this approach. First we must measure variability in a population since all individuals are not identical copies of one another. Second, we must define how much chance we are willing to live with in making decisions.

The first issue is how to report the beneficial or adverse results of a study when we know that all individuals do not respond the same. We can never, nor would we want to, experiment on *every* human. Therefore, studies are conducted on a portion, or sample, of the population. We hope that the people selected are representative of the larger population. Since a number of people are studied and all never respond identically, a simple average response is not appropriate. Variability must be accounted for and a range described. By convention, we usually make statements about how 95% of the population in the study responded. We then use probability and extend this estimate so that it most likely is true for 95% of the population. If the effect of our drug or chemical being tested makes a change in response that is outside of what we think is normal for 95% of the population, we assume that there is an effect present.

Let us take a simple example. Assume you have 100 people ranging in weight from 125 to 225 pounds, with the average weight being 175 pounds. You need some additional information to describe the spread of weights in these groups. Generally a population will be characterized by an average or "mean" with gradually decreasing numbers of people of different weights on either side described by a so-called "normal" or "bell-shaped" distribution. It is the default standard descriptor of a population. However, you could have a situation with 100 people actually weighing 175 pounds or two groups of 50 people each weighing 125 and 225 pounds. The average of these three groups is still 175 pounds; however, the spread of weights in each scenario is very different, as shown in Figure 1.1.

Some estimate of variability must be calculated to measure just how different the individuals are from the ideal mean—in other words, a way to see how wide the bell really is. To achieve this, statisticians describe a population as the mean plus or minus some factor that accounts for 95% of the population. The first group above could be described as weighing 175 pounds, with 95% of the people weighing between 150 to 200 pounds. This descriptor holds

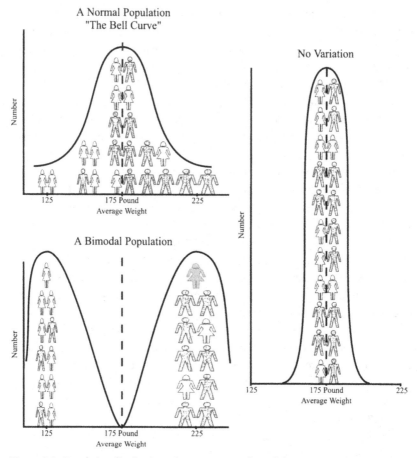

Figure 1.1. Population distributions: the importance of variability.

for 95 people in our group, with 5 people being so-called outliers since their weights fall beyond this range (for example, the 125 and 225 pound people). In the bimodal population depicted in the figure (the two groups of 50), the estimate of variability will be large, while for the group consisting of identical people, the variability would be zero. In fact, not a single individual in the bimodal population actually weighs the weight of the average population (there is no 175 pound person).

In reality, both of these cases are unlikely but not impossible. Our "group" could have been two classes of fifth- and eleventh-grade students that would have been a bimodal population very similar to that described above. In most cases, the magnitude of the variability is intermediate between these extremes for most normally distributed populations. Just as we can use weight as a descriptor, we can use other measurements to describe disease or toxicity induced by drugs or pesticides. The statistical principles are the same for all.

Now in the above example, we had actually measured these weights and were not guessing. We were able to measure all of the individuals in this group. If we assume that this "sample" group is similar to the whole population, we could now use some simple statistics to extend this to an estimate—an inference—of how the whole population might appear. *This ability to leap from a representative sample to the whole population is the second cornerstone of statistics.* The statistics may get more complicated, but we must use some measure of probability to take our sample range and project it onto the population. One can easily appreciate how important it is that this sample group closely resemble the whole population if any of the studies we are to conduct will be predictive of what happens in the whole population. Unfortunately, this aspect of study design is often neglected when the results of the study are reported.

Let's use a political example. If one were polling to determine how an election might turn out, wouldn't you want to know if the sample population was balanced or contained only members of one party? Which party were they in? The results of even a lopsided poll might still be useful, but you must know the makeup of the sample. Unfortunately, for many of the food safety and health studies discussed throughout this book, there are not easy identifiers such as one's political party to determine if the sample is balanced and thus representative of the whole population. Best's book *Damned Lies and Statistics* contains a number of other cases such as these where statistics were in his words "mutated" to make a specific point.

Just as our measured description was a range, our prediction will also be a range called a "confidence interval." It means that we have "confidence" that this interval will describe 95% of the population. If we treat one group with a drug, and not another, confidence intervals for our two groups should not overlap if the drug truly has an effect. In fact, this is one way that such trials are actually conducted.

This is similar to the approach used by the political pollsters above when they sample the population using a poll to predict the results of an upcoming election. The poll results are reported as the mean predicted votes plus a "margin of error" which defines the confidence interval. The larger and more diverse the poll, as well as the closer to election day, the narrower the confidence interval and more precise the prediction. The election is the real measurement of the population, and then only a single number (the actual votes cast) is needed to describe the results. There is no uncertainty when the entire population is sampled. Unlike in politics, this option is not available in toxicology.

As you can see, part of our problem is determining how much confidence we have, that is, how much chance are we willing to accept? Why did we select 95% and not 99%? To accomplish this, some level of confidence must be selected below which the results could be due to chance alone. We want to make sure our drug actually produces an effect and that the effect is not just present because, by chance, we happened to sample an extreme end of the range. The generally accepted cutoff point for the probability of the results being due to chance is less than a 1 in 20—that is, 5 out of 100, or approximately 5%. In statistical lingo, this reads as a "statistically significant finding with a $P < 0.05$" where P is the probability that chance caused the results. *This does not mean that one is certain that the effect was caused by the drug or chemical studied.* One is just stating that, assuming that the sample studied was representative of the general population, it is unlikely that the effect would occur by chance in the population.

How is this applied in the real world? If one were testing antibiotics or anticancer drugs, two situations with which I have a great deal of personal experience, I would not get excited until I could prove, using a variety of statistical tests, that my experimental results were most likely due to the treatment being tested and not just a chance occurrence. Thus for whatever endpoint selected, I would only be happy if it had a less than 1-in-20 chance of being caused by chance alone. In fact, I would only "write home to mom" when there was a 1-in-100 chance ($P < 0.01$) that my results were not due to chance.

Why am I concerned with this? If I take the sample group of people from Figure 1.1 and give them all a dose of drugs or chemicals based on their average weight of 175 pounds, there will be as many people "overdosed" as "underdosed." But what if I sample only 5 people and they are all from the lighter end of the range? What happens if they were in a bimodal population and only the 125-pound people were included? All would be overdosed and toxicity would be associated with this drug purely on the basis of a biased sample. Although such a situation might be easy to detect because of obvious differences in body size, other more subtle factors (e.g., genetic predisposition, smoking, diet, unprotected extramarital sex) that would predispose a person to a specific disease might be easily missed. The study group must be identical to the population if the prediction is to be trusted!

A likely scenario would be a patient who is spontaneously cured. In a small study with perhaps only 10 patients to a group, 1 patient in the control group may cure, completely independent of any treatment being tested. If this tendency to "spontaneously cure" is present in all patients treated, then 1 in 10 treated patients will get better independent of the drug. Statistics is the science that offers guidance as to when a treatment or any outcome is significantly better than doing nothing. If the disease being treated has a very high rate of spontaneous cure, then any so-called treatment would be deemed effective if this background normal rate of spontaneous cure is not accounted for. In designing a well-controlled study, I must be assured that the success of my treatment is not due to chance alone and that the drug or treatment being studied is significantly better than doing nothing. The untreated control and treated confidence intervals should not overlap.

Most scientists, trained to be skeptical of everything, go further and repeat a positive study just to be sure that "lady luck" is not the real cause of the results. If this drug were then to be approved by the Food and Drug Administration (FDA) as a treatment, this process of reproducing promising results in an increasingly complex gauntlet of ever more challenging cases would have to continue. At each step, the chance rule is applied and results must have at least a P<0.05 level of significance. In reality, pharmaceutical companies will design studies with even more power (p<0.01 or even p<0.001) to be sure that the effect studied is truly due to the drug in use. This guarantees that when a drug finally makes it on the market, it is better than doing nothing. This obviously is important, especially when treating life-threatening diseases.

Considering what this 1-in-20 chance really means is important. It is the probability that the study results are due to the effect being tested and not due to chance. If 20 studies were repeated, chance alone would dictate that 1 study might indicate an effect. This would be termed a "false positive." One must include sufficient subjects to convince the researcher that the patients are getting better because of the treatment or, in a chemical toxicity study, that the effect seen is actually related to chemical exposure. The smaller the effects that the study is trying to show, the more people are needed to prove it.

Most readers have some practical experience with the application of probability in the arena of gambling. When someone flips a coin, there is always a 50/50 chance that a head or tail will result. When you flip the coin, you accept this. You also know that throwing 10 heads in a row is unlikely unless the coin is unbalanced (diseased). The final interpretation of most of the efficacy and safety studies confronted in this book, unfortunately, are governed by the same laws of probability as a coin toss or winning a lottery. It boils down to the simple question: When is an event—cure of AIDS or cancer, development of cancer from a pesticide residue—due to the drug or chemical, and when is the outcome just a matter of good or bad luck— picking all your subjects from one end of the population range? When the effect is large, the interpretation is obvious. However, unfortunately for us, most often this is not the case.

Application of the statistical approach thus results in the rejection of some useful drugs. For example, assume that Drug X completely cures cancer in only 1 of 50 patients treated. In a drug trial, it would be hard to separate this 1-in-50 success from a background of spontaneous cures. Drug X would probably be rejected as an effective therapy. Why? Because one must look at the other side of the coin. In this same study, Drug X would not work in 49 of 50 patients treated! It is this constant application of statistics that has resulted in assuring that new treatments being developed, even in the initial stages of research, will significantly improve the patient's chances of getting better.

For our Drug X, if the results were always repeatable, I would venture that

someone would look into the background of the 1 in 50 that always responded and determine if there were some subgroup of individuals (e.g., those with
a certain type of cancer) that would benefit from using Drug X. This would be
especially true if the disease were one such as AIDS where any effective treatment would be considered a breakthrough! In fact, in the case of AIDS where
spontaneous recovery just doesn't happen, the chances would be high that
such a drug would be detected and used. Individuals with the specific attributes associated with a cure would then be selected in a new trial if they could
be identified before treatment began. The efficacy in this sensitive subgroup
would then be tested again.

However, if the drug cured the common cold, it is unlikely that such a poor
drug candidate would ever be detected. Of course, because 1 in 50 patients
who take the drug get better, and approximately 10 million people could
obtain the drug or chemical before approval, there would be 200,000 cured
people. Anecdotal and clinical reports would pick this up and suggest a breakthrough. The unfortunate truth is that it would not help the other 9,800,000
people and might even prevent these individuals from seeking a more effective therapy. Such anecdotal evidence of causation is often reported and not
easily discounted.

The reader should now ask, what about the patients who improve "by
chance" in approximately 1-in-20 times, the acceptable cutoff applied in most
studies? There are as many answers to this intriguing question as there are
experiments and diseases to treat. For many infectious diseases studied, often
a patient's own immune status may eliminate the infection, and the patient
will be cured. Recall that few infectious diseases in the past were uniformly
(e.g. 100%) fatal and that even a few hardy individuals survived the bubonic
plague in medieval Europe or smallpox in this country. These survivors would
fall in the 1-in-20 chance group of a modern study. However, without effective
antibiotics, the other 19 would die! Using today's principles of sound experimental design, drugs would only be developed which increased the chances of
survival significantly greater than chance. This is the best approach. Reverting
to a level of spontaneous cures would throw medicine, along with the bubonic plague, back to the dark ages.

Another explanation of this effect is the well-known placebo effect, whereby it has been shown time and time again that if a patient is taking a drug that
he or she believes is beneficial, in a significant number of cases a cure will
result. That placebos (e.g., sugar-pills) will cure some devastating diseases has
been proven beyond doubt. Thus a drug is always tested against a so-called
placebo or control to insure that the drug is better than these inactive ingredients. This is why most clinical trials are "blinded" to the patient and doctors
involved. The control group's medication is made to look identical with the
real drug. In many cases, the control is not a placebo but rather the most effec-

tive therapy available at the time of the trial since the goals are usually to improve on what already exists. At certain preselected times, the results of the trial are compared and if the new drug is clearly superior at a $P<0.05$ level, the trial may then be terminated. However, even if the drug looks promising but chance has not quite been ruled out, the trial will continue and the drug will not be approved until clear proof of efficacy is shown.

A similar effect is probably responsible for much of the anxiety that surrounds the potential adverse effects of consuming pesticide residues on fruits and vegetables—the primary topic of this book. Many types of diseases will spontaneously occur and affect small segments of our population (e.g. at the ends of the ranges) even if pesticides never existed. However, if these individuals were also exposed to pesticides, one can understand how a connection may be made between disease and chemical exposure. One problem with making such associations is that all of the other people exposed to dietary pesticides do not experience adverse effects and, in fact, may even see a significant health benefit from consuming fruits and vegetables, even if pesticide residues are present. Another problem in this area relates to the placebo effect itself. If an individual *believes* that pesticides will cause disease, then just like the sugar-pill that cured a disease, pesticide exposure may actually cause disease as a result of this belief. This is a dangerous misconception and may be responsible for individuals becoming sick solely due to the belief that chemicals may cause a specific disease!

Most will agree that the consistent application of sound statistical principles to drug development is to their collective long-term benefit. It guarantees that drugs developed actually do what they are intended to do. However, the application of the same principles to chemical toxicology and risk assessment, although covering the identical types of biological phenomenon, is not as widely understood. It is troublesome when such hard statistical rules (such as the 1-in-20 chance protector), which are well understood and accepted for drug efficacy, are now applied to safety of drugs or environmental chemicals and interpreted differently. In this case, the 1 in 20 responders are not positive "spontaneous" cures but instead are negative "spontaneous" diseases! The science has not changed, but the human value attached to the outcome is very different!

It is human interpretation that gets risk assessment in trouble. There are two situations where the application of hard statistical rules, which are so useful in assessing efficacy, become problematic when applied to drug- or chemical-induced toxicity. Those situations are when promising drugs are rejected because of adverse effects in a small number of patients and when serious spontaneous diseases are wrongly attributed to a presence of a chemical in the environment. It seems that in these cases the rules of statistics and sound experimental design are repealed because of overwhelming emotion and compassion for the victims of devastating disease.

An example of the first case is when a drug that passes the 1-in-20 chance rule as being effective against the disease in question may also produce a serious adverse effect in say 1 in 100 patients. Do you approve the drug? I can assure you that if you are one of the 99 out of 100 patients that would be cured with no side effect that you would argue to approve the drug. People argue this all the time and fault the FDA for being too stringent and conservative. However, if you turn out to be the sensitive individual (you cannot know this prior to treatment) and even if you are cured of the underlying disease, you may feel differently if you suffer the side effect. You might be upset and could conceivably sue the drug company, doctor, and/or hospital asking, "Why didn't they use a different drug without this effect?" Your lawyer might even dig into the medical literature and find individuals in the population that were spontaneously cured to show that a drug may not have been needed at all. Because of this threat, the above numbers become too "risky," and instead the probability of serious adverse effects is made to be much lower, approximately 1 in 1000 or 1 in 10,000. Although this protects some patients, many effective therapies that would benefit the majority are rejected because of the remote chance of adverse effects in the sensitive few.

Unfortunately this judgment scenario was being played out as I write this chapter when anthrax was detected in letters in Washington, DC. Before this incident, the military's vaccine against anthrax was being attacked due to the possibility of serious adverse effects when the risk of exposure was low. Today, with anthrax exposure actually being detected, the risks do not seem as important. The same benefit/risk decision can occur with taking antibiotic after potential exposure. What will an individual who does not contract anthrax after taking preventative measures do if a serious adverse effect occurs?

Value judgments, the severity of the underlying disease, and adverse reaction thus all factor in and serve as the basis of the science of risk assessment. For example, if you have AIDS, you are willing to tolerate many serious adverse effects to live a few more months. The same holds for patients in the terminal stages of a devastating cancer. The result is that the drugs used to treat such grave diseases are often approved in spite of serious side effects, as any one who has experienced chemotherapy knows. However, if the disease is not lethal—for example the common cold or a weight-loss aid—then the tolerance for side effects is greatly lowered. In most of these cases, the incidence of side effects drops below the level of chance, and herein is the cause of what I believe to be the root of many of our problems and fears of chemically induced disease. That is, if a person taking a drug experiences an adverse effect that may remotely be associated with some action of a drug, the drug is blamed. However, analogous to some people "spontaneously" getting better or being "cured" because of the placebo effect, others taking a drug unfortunately get sick for totally unrelated reasons or due to a "reverse-placebo" effect. They would be members of that chance population that would experience a side effect even when drug or chemical exposure did not occur. The same situation may hold true for some of the purported adverse health effects of low-level pesticide exposure. Although the cause is not biological, the illness is still very real.

Pharmaceutical companies and the FDA are fully aware of this phenomenon and thus have instituted the long and tortuous path for drug approval that is used in the United States. The focus of this protracted approval process (up to a decade) is to assure that any drug that finally makes it into widespread use is safe and that adverse effects that do occur are truly chance and will not be associated with drug administration. In contrast, some adverse effects are very mild compared to the symptoms of the disease being treated and thus are acceptable. Post-approval monitoring systems are also instituted whereby any adverse effects experienced by individuals using the drug are reported so that the manufacturer can assure that it is not a result of the drug. Such systems recently removed approved weight-loss, diabetic, and cholesterol-lowering drugs from the market. In the case of serious diseases—for example AIDS, cancer, Alzheimer's disease, Lou Gehrig's disease (amyotrophic lateral sclerosis)—these requirements are often relaxed and approval is granted under the umbrella of the so-called abbreviated orphan drug laws that allow accelerated approval.

Even in cases of serious diseases, the requirements for safety are not completely abandoned thus assuring that the most effective *and* safe drugs are developed. This requirement was personally "driven home" when the spouse of an early mentor of mine recently died from Lou Gehrig's disease. While she was in the terminal phases of this disease, my friend contacted me, frantically searching for any cure. The best hope occurred when she was entered into a clinical trial for the most promising drug candidate being developed at the time. Unfortunately, the trial was terminated early because of the occurrence of a serious side effect in a number of patients. The drug was withdrawn and the trial ended. My colleague's beloved wife has since passed away. Was this the right decision for her?

The reader should realize that in some cases, separating the adverse effect of a drug from the underlying disease may be difficult at best. For example, I have worked with drugs whose adverse effect is to damage the kidney, yet they must be given to patients who have kidney disease to begin with. For one anticancer drug (cisplatin), the solution was not to use the drug if renal disease was present. Thus patients whose tumors might have regressed with this drug were not treated because of the inability to avoid worsening of the renal disease. The ultimate result was that new drugs were developed which avoided the renal toxicity. For another class of drugs, the aminoglycoside antibiotics, strategies had to be developed to treat the underlying bacterial disease while minimizing damage to the kidney. Since the patients treated with these drugs had few alternatives due to the severity of the bacterial disease, these approaches were tolerated and are still used today.

However, for both of these drugs, the problem has always been the same. If a patient's kidney function decreased when the drug was used, how could you separate what would have occurred naturally due to the underlying disease from effects caused by the drug against the kidney? A similar situation recently occurred when a drug used to treat hepatitis had also caused severe liver toxicity. In these cases of serious disease, extensive patient monitoring occurs, and the data is provided to help separate the two. However, what happens when drugs are used that have a much lower incidence of adverse effects but, nonetheless, could be confused with underlying disease?

Another example from my own experience relates to trying to design laboratory animal studies that would determine why some individuals are particularly sensitive to kidney toxicity. Considering a number of the studies discussed above and numerous reports in the literature, there always appeared to be a small percentage of animals that would exhibit an exaggerated response to aminoglycoside drugs. If these individuals appeared in the treatment groups, then the treatment would be statistically associated with increased toxicity. However, if they were in the control group, they would confound (a statistical term for confuse) the experiment and actually generate misleading results. This was happening in studies reported in the literature—sometimes a study would indicate a protective effect of giving some new drug with the kidney-toxic aminoglycoside and, in other cases, no effect. This lack of reproducibility is disconcerting and causes one to reevaluate the situation. Based upon the analysis of all available data, we thought that the problem might have been a sensitive subpopulation of rats—that is, some individuals were prone to getting kidney disease—and so we decided to investigate further. At this point I believe it is instructive to review the process of designing this experiment, as we need to remember that the assumptions made up front, before anyone ever starts any real experiments, will often dictate both the outcome and the interpretation of results.

We were interested in determining what makes some people and rats sensitive to drugs. Since one cannot do all of these studies in people, an appropriate animal model is selected. For reasons too numerous to count, the choice of animal for these antibiotic studies came down to two strains of rats, both stalwarts of the biomedical researcher: the Sprague-Dawley and Fischer 344 white laboratory rats. The difference between these two strains is that the first is randomly outbred and has genetic diversity, much like the general human population. The Fischer is highly inbred (that is, incestuous) and might not be as representative. We selected Sprague-Dawley rats because they were used in many of the earlier studies. When sophisticated mathematical techniques (pharmacokinetics) are used to extrapolate (to project) drug doses across animal species ranging from humans to mice, rats, dogs, and even elephants, we also knew that the Sprague-Dawley rat is a good choice because the data has historically been projected easily across all species.

To determine if some rats were sensitive, we elected to give 99 rats the identical dose of the aminoglycoside, gentamicin. We then monitored them for signs of kidney damage by taking blood samples and examining their kidneys after the trial was over. To some individuals who believe that this use of animals is not humane, I must stress that when these studies were conducted a decade ago, humans being treated with these drugs were either dying (or going on dialysis) from kidney damage or dying from their underlying infections. These studies were funded by the National Institutes of Health (NIH) who judged them to be scientifically justified and the animal use humane and warranted.

What did these studies show? In a nutshell, 12 out of 99 animals—approximately 1 in 8—had a significantly exaggerated response to gentamicin at such a level that if they appeared in only one treatment group of a small study, this enhanced response would have been erroneously associated with any treatment. They statistically behaved as a bimodal population (12 versus 87 members) with two different "mean" treatment responses. The extensive pretreatment clinical testing and examinations we conducted did not clearly point to a cause for the sensitivity. In this study, we *knew* that all rats were treated identically and thus there was no effect of other drugs (since none were given), or differences in housing or diet since they were identical. What was even more enlightening was that every eighth rat did not show the response. We might see 20 rats with no response and then suddenly a run of 4 sensitive individuals would occur (a cluster as defined by a statistician). Our experimental design guaranteed that there was no connection between the sequence in which these animals were observed and the response, making the "cluster" purely random. The sensitive animals were not littermates—they all came from different cages—and they ate the same diet as the normal animals.

Such data is often not available in the field for patterns of disease in human populations, and thus spurious associations may be made when a cluster is encountered. In reality, it often is only due to chance, an event analogous to coin flipping and getting four heads in a row.

Why have I discussed antibiotic rat studies in a book on chemical and pesticide food safety? Because this study clearly shows how a "background" or "spontaneous" disease, in this case sensitivity to a kidney-damaging chemical, can arise in normal groups of animals. If one looks at this sensitive subpopulation of Sprague-Dawley rats, they appear by some parameters to be similar to the inbred Fischer 344 rat not selected for study, who also show an increased sensitivity to these drugs. It is possible that a full genetic analysis of these rats would determine a cause. This phenomenon is widely observed in biology and medicine for any number of drugs or diseases; however, it is often forgotten or deliberately ignored when attempts are made to link very low *"background"* incidences of disease to a drug or environmental chemical exposure. For example, many of the laboratory animal studies which once suggested that high doses of pesticides cause cancer, used highly inbred strains of mice that are inherently sensitive to cancer-causing chemicals. In the case of cancer, the reason for this sensitivity may be the presence of oncogenes (genes implicated in causing certain cancers but not known at the time of the studies) in these mice. The use of a highly sensitive mouse to detect these effects is often forgotten when results are reported.

An excellent example illustrating the dilemma of separating drug effect from background disease was the removal of the antinausea drug Bendectin® from the market. This drug was specifically targeted to prevent morning sickness in pregnant women. However, in some individuals taking the drug, birth defects occurred and numerous lawsuits were filed against the manufacturer. The FDA launched its own investigations, and additional clinical studies were conducted looking for a link between Bendectin® and birth defects. After FDA and multiple scientific panels examined all of the evidence, no association was found! Only a few studies with serious design flaws indicated a weak connection.

In these types of epidemiological studies, a relative risk is often calculated which compares the risk of getting a disease with the drug (in this case a birth defect) against the "spontaneous" rate seen in the population not taking the drug.

$$\text{Risk Ratio} = \frac{\% \text{ of affected (e.g., birth defects) in treatment (e.g., drug) group}}{\% \text{ of affected in control group or background population}}$$

This ratio or proportion is computed using ranges of affected individuals. A result of 1.0 would indicate no difference from the control population, meaning that confidence intervals of control and treated groups overlap. In other words, if the calculated ratio is 1.0, then there is no statistical difference between the incidence in the control and treated groups. If the ratio was > 1.0, then there is some risk of taking the drug.

After all the Bendectin® studies were examined and the statistical differences mentioned earlier accounted for, the relative risk of birth defects from taking the drug was 0.89, with a confidence interval from 0.76 to 1.04. This means that for most women, there was almost a statistically significant protective effect since the ratio was generally < 1.0. However, for most patients, the risk was equivalent to the control group, and thus there was no drug effect on this endpoint.

What happened? The unfortunate truth of the matter is apparently that the "spontaneous" or "background" rate for birth defects in the American population hovers around 3% or 3 birth defects in 100 live births. So even if a pregnant woman takes no drug, there will still be 3 out of 100 births with birth defects. These are obviously not related to taking a nonexistent drug (or being exposed to a pesticide residue on an apple, but I am getting ahead of myself). However, 30 million women took Bendectin® while they were pregnant, and one would predict that 3% of these, or 900,000 would have birth defects not related to the drug. You can well understand why a woman taking this drug who delivered a baby with a birth defect would blame the drug for this catastrophic outcome. However, even with removal of the drug from the population because of litigation, we still have birth defects in 3 out of 100 births, some of which, it could be argued from the data above, could have even been prevented by use of the drug. Note that I am not arguing for this side of the coin. However, severe morning sickness and nausea could conceivably cause medical conditions (e.g., dehydration, stress) predisposing to adverse effects. Remember that the drug was originally approved because of its beneficial effects on preventing morning sickness. Was this the correct decision?

One could complicate this analysis even further by considering why morning sickness even exists in the first place. Based upon an evolutionary perspective on medicine, some workers have postulated that the nausea and altered eating patterns associated with pregnancy that discourage consumption of certain strong-tasting foods is an evolutionary adaptation to protect the vulnerable fetus from being exposed to *natural* teratogens (chemicals that cause birth defects) in the diet. Since this is exaggerated during the first trimester, maximum protection would be provided when the developing fetus is at greatest risk. Such proponents would have even argued that if the Bendectin® trials had shown an *increase* in birth defects, the reason may have been due to this blunting of the protective effects of morning sickness rather than a direct drug effect. Again, I am not writing to argue for this perspective but mention it to illustrate that the data has no meaning unless interpreted according to a specific hypothesis. As the data suggests, if anything, Bendectin® may have protected against birth defects, shedding doubt on this hypothesis which may have been pertinent if mothers were still eating Stone Age diets!

Finally, this is also an opportune point to introduce another concept in epidemiology which has recently taken on increased importance. This is the question: How large or significant should a risk be for society to take heed and implement action? As can be seen from the Bendectin® example above, the potential risk factors were hovering around 1.0. A recent review of this problem in *Science* magazine indicates that epidemiologists themselves suggest concern be warranted only when *multiple* studies indicate risk factors *greater* than 3.0 or 4.0! If risks aren't this high (e.g., Bendictin®), then all we are doing is increasing anxiety and confusing everyone. I would add that we might even be doing a great harm to public health by diverting scarce attention and resources away from the real problems. There are so many false alarms that when the real alarm should be sounded, no one listens. For a better perspective on this, I would suggest reading Aesop's *The Boy Who Cried Wolf!*

This problem may be even more rampant in the field of environmental epidemiology where exposure to any drug or chemical is rarely measured. In fact, some leading epidemiologists suggest that this field has only been accurate when very high-risk factors are present, such as smoking, alcohol, ionizing radiation, asbestos, a few drugs, and some viruses. Repeated studies have shown elevated risks to certain populations. Even for these, there are often significant interactions between the dominant risk factors of smoking and chronic alcoholism. However, many of the other "environmental scares" identified in the past few decades have been based on limited studies with risk factors of under 2.0, which in subsequent follow-up are found lacking. One could even argue that for a study to be significant, the lowest level of the statistical interval of the study (all 95% of respondents) should have a risk factor *greater* than 3 or 4. These suggestions are from epidemiologists themselves. The public should take heed, as taking inappropriate action may be detrimental to all of us.

As can be seen from this brief overview of the application of statistics to certain aspects of medicine, the science is well defined; however, the rules for interpretation are not clear. Data will be presented in this book that fruit and vegetable consumption, with trace levels of pesticides, actually *decreases* the risk of getting cancer and *increases* lifespan.

There will always be spontaneous disease in the population whose cause is not attributable to the drug or chemical being studied. In a drug trial, this is the placebo effect and results in a cure. In a toxicology study, this is an individual who contracts a disease that in no way is caused by the chemical in question or has a true psychosomatic condition brought upon by the belief that a chemical is causing harm. All scientific studies must have controls for this background incidence of disease and take precaution not to associate it with spurious environmental factors to which the control groups are also exposed. Finally remember that death is inevitable and, in every population, people will die. We cannot blame chemicals for this finality no matter how much we would like an explanation. With this background, let us now do some real toxicology.

> Pale Death, with impartial step, knocks at the poor man's cottage and the palaces of kings.
>
> (Horace, *Odes, 1, 4*)

2

Dose Makes the Difference

A major source of confusion contributing to the debate on the safety of pesticide residues in food, and low-level exposure to environmental chemicals, is the relationship between a chemical exposure or dose and the observed effect. *Some* pesticides to which we are exposed in our fruit and vegetables are capable of causing harm to humans *if given in high enough quantities.* This is usually limited to occupational exposure where high toxic doses may actually be encountered. If the need for statistical analysis to rule out chance effects is the first scientific tenet that forms the cornerstone of our modern science-based medicine, the dose-response relationship is the second.

The effect of a chemical on a living system is a function of how much is present in the organism. Dose is a measure of the amount of chemical that is available for effect. Every beginning student of pharmacology, toxicology, and medicine has been exposed to the following quotation from Paracelsus (1493–1541), whom many consider to be the father of toxicology and godfather of modern chemotherapy:

> What is there that is not poison? All things are poison and nothing (is) without poison. Solely the dose determines that a thing is not a poison.

It is important to seriously ponder what this statement means because its application to risk assessment is the basis for most of society's generally unfounded concern with pesticides and chemicals. To illustrate this, it is best to first apply this rubric to circumstances that are familiar to everyone. As in Chapter 1, the science is the same across all chemicals; it is the interpretation clouded by a fear of the unknown that affects our ability to make a clear judgment.

Let's start off with a familiar chemical that is known to cause serious toxicity in laboratory animals and humans. This compound may result in both acute and chronic poisoning when given to man and animals. Acute poisoning is that which results from a short-term (few doses) exposure to high levels of the chemical in question; whereas chronic effects relate to slower developing changes that are a result of long-term exposure to lower doses of the chemical in question. High-dose exposure to this chemical has been well documented in acute toxicity studies to cause vomiting, drowsiness, headaches, vertigo, abdominal pain, diarrhea, and, in children, even bulging of the fontanelle—that soft spot on a baby's head that for generations mothers have warned others not to touch when holding your new infant. Chronic exposure to lower doses in children results in abnormalities of bone growth due to damage to the cartilage. In adults symptoms include anorexia; headaches; hair loss; muscle soreness after exercise; blurred vision; dry, flaking, and itchy skin; nose bleeds; anemia; and enlargement of the liver and spleen. To make matters worse, this nasty compound and its derivatives are also teratogenic (cause birth defects) in most animals studied (dogs, guinea pigs, hamsters, mice, rabbits, rats, and swine) and are suggested to do the same in humans. Looking at this hard-fast evidence, reproduced in numerous studies and supported by most scientific and medical authorities, one might be concerned if there was widespread exposure to humans.

In fact, there is widespread exposure to this chemical since it is found in many foods we eat. Milk is even fortified with it! Oops, I forgot to tell you that this nasty chemical is actually vitamin A. It, and the related natural carotenoids, are compounds that at lower doses are widely acknowledged to be beneficial to man. Their derivatives (retinol, retinin A) are used therapeutically to help prevent cancer. However, at high doses they are known to produce adverse effects. This is a clear-cut example showing that if only high-dose data is used, which for vitamin A is about 10–20 times the maximum recommended daily dose, toxicity may be the result. Also, these are not theoretical risks since these syndromes have been well documented in humans. Chapter 5 will present a litany of other natural substituents of our foods that in high-dose laboratory studies have been shown to be teratogenic and even carcinogenic, yet long-term human data suggests that consumption of food containing them is actually beneficial to our health. In fact, many natural chemicals found in vegetables may fall in this group.

Back to how the vitamin A data can be interpreted. This is a compound that clearly is dangerous at high doses and beneficial at lower doses. It is an essential vitamin, so its complete absence in our food would result in a malnutrition disorder that constantly plagued our ancestors. Thus Paracelsus's

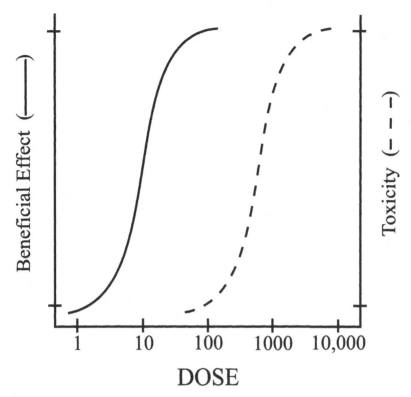

Figure 2.1. Dose response curve.

rubric holds well—the dose does not only make the poison but could be extended to also make the medicine. The problem comes when one extrapolates the data and changes the endpoints evaluated. This is illustrated in Figure 2.1, in which both adverse and beneficial dose-response curves for a compound such as vitamin A are plotted.

If only high-dose laboratory animal and human data which result in adverse effects are used, and this data is extrapolated (extended or projected) to lower doses, then one would predict that there could be adverse effects even at very low intake of the compound. However, we have data covering generations of people exposed to these low doses that show no adverse effect and actually significant benefit. (Note, we would not have this data if this were a new chemical because it would never have made it onto the market in the first place!) We also know from extensive studies, there are "thresholds" for vitamin A effects. A threshold is an amount, above which the vitamin does one thing, and below which it does another or nothing at all. The problem with dose extrapolations to very low levels is that we often are not as fortunate and do not know if and where thresholds occur. We often blindly assume that there are none.

The predicted risk to humans based on these types of studies is completely dependent upon the underlying assumptions used to make the prediction. If we only have the high-dose data, and even if we determine in laboratory animal studies that there is a dose-response curve and a threshold exists, we often must use a straight-line (linear) approach from this high-dose data since it is the only reliable and legally accepted data we have. However, other models are possible if more data are available and certain assumptions are made. Whatever we do, we are essentially guessing since no human data is collected at these lower doses. This process is depicted in Figure 2.2.

Because the no-effect level can often be only statistically proven to the level of the background noise (recall our discussion in Chapter 1 about spontaneous disease level), it is impossible to statistically prove a zero- or no-effect dose. In fact, as will be touched on later, subtoxic chemical exposure may be beneficial for many chemicals as homeostatic compensatory mechanisms are induced which have a beneficial, not adverse, biological effect.

This basic limitation of the scientific method is often misinterpreted when findings from risk assessment studies are reported in the popular press. A scientific conclusion may be stated in terms that "it is highly unlikely that" or

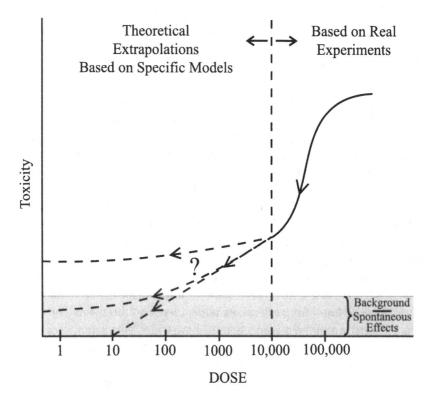

Figure 2.2. The pitfalls of extrapolation.

"there is no data to suggest that" some adverse effect will be seen from a chemical at a low dose. However, one must always add the caveat that this "no-effect" level cannot be proven due to the limitations of extrapolation and background interference. Journalists may take this caveat and conclude that the study "cannot prove a no-effect level for this compound." This is simply wrong.

There is not a conceivable laboratory animal experiment that would prove that a chemical does not cause cancer in 1 out of 1 million or 1 out of 10 million people exposed to it. You could not prove this with vitamin A because some individuals who take this supplement for life will die of cancer and some mothers who take it during pregnancy will have children born with birth defects. Of course, this will also happen in the general population not exposed to the chemical. Just because these compounds are capable of causing toxicity at high doses, does not mean that they will cause harm at lower levels!

I admit that some may consider vitamin A as an extreme example of this effect. We now know that at high doses, vitamin A interacts with the membranes of our cells differently than at lower doses. However, similar mechanisms are true for many types of drugs and chemicals and are completely ignored when some claim that a chemical implicated in high-dose animal studies should be banned from the planet because of fear of what it might do at lower doses. The potential loss of beneficial health effects by using this kind of strategy is frightening.

What other vitamins and minerals share this problem of being toxic at high doses, beneficial at moderate doses, and even essential at trace or residue levels? One is vitamin D, which if dosed high enough in animals and man, produces toxicity associated with the resulting high blood levels of calcium. The signs of poisoning are formation of calcium deposits in nonskeletal tissues, hypertension, kidney disease, and cardiac failure. Similarly, high doses of vitamin E and K may also produce toxicity. Ditto for some minerals (e.g., selenium) and even some amino acids (for example methionine, isoleucine and threonine). The adverse effects of these essential nutrients only occur at very high doses and are not associated with common dietary intake levels. I take vitamin supplements because their proven beneficial effects far outweigh any theoretical risk of toxicity.

However, this is precisely the point of this entire book—just because a chemical is capable of producing an adverse effect at extremely high doses, and in some cases almost suffocating ones, it does not follow that low doses will produce adverse effects. For the pesticides, I will only be arguing that at low doses (below FDA acceptable levels), there are minimal adverse biological effects. I hesitate to say that if these vitamins had to pass studies mandated by today's regulatory agencies, they would not be approved!

As an aside at this point in the discussion of dose-response relationships, it is important to point out the existence of a recent movement describing a new type of dose-response curve that is being observed for many chemicals— the U-shaped dose-response phenomenon. Unlike the extrapolations depicted in Figure 2.2, at low concentrations, the chemical actually shows fewer adverse effects than at higher concentrations and even may demonstrate a beneficial effect. The U-shaped term comes from the description of a dose-response plot of beneficial effects which looks like a U. The mechanism of this effect, also termed *hormesis* (Greek for *excite*), is not completely known, but it is believed secondary to the effect of low chemical exposures to induce metabolic responses beneficial to the organism. Hormesis will be revisited in Chapter 4.

Why do we not worry about these common substances such as vitamins that obviously have adverse effects at high doses? The answer is because they are common and we have experience with them. We are familiar with their names as we have been exposed to them since birth. The toxicological principles are the same with synthetic chemicals; however, we tend to shut off our logic and common sense and let our fears run wild. This may be fine for the individual, but as a member of society at large, I do not want my health impaired because of unfounded fears of toxicity.

Let us move to an area that in some people's minds may be more pertinent to the concerns of synthetic chemical toxicity. This is the application of the dose-response principle to the development and marketing of drugs. Most people know that the dose is an important factor to consider when taking any kind of medication. This holds for the simple drugs such as aspirin to more powerful tranquilizers, sedatives, and antibiotics. It is common knowledge that if these useful compounds are taken at too high of a dose, serious problems and even death may result. This, unfortunately, is one reason that suicide by overdose of certain medications is a far too common occurrence in our society. It is also commonly appreciated, and accepted as a necessary evil, that there are serious side effects seen when high-dose chemotherapy is used to treat life-threatening cancer. However, for most drugs on the market used to treat less serious illnesses, no problems arise if one uses proper dosages. People generally do this and have no anxiety about consuming chemicals that at higher doses would kill them. Common sense also tells us to ignore high-dose data since we have always been taught that too much of a good thing may be bad for us. In contrast, we have often codified the exact opposite into our regulations concerning chemical exposure.

The *Merriam Webster's Collegiate Dictionary* defines toxicology as "a science that deals with poisons and their effect and with the problems involved (as clinical, industrial, or legal)." The science of toxicology has been instrumental in developing the animal and early human trials to develop new pharmaceuticals. The dose-response relationship is the defining principle of this science.

We trust these decisions when they apply to our synthetic medications, so why do we ignore the same principles when they apply to chemicals having much greater margins of safety?

The approach to drug safety testing is well developed. Initial studies are conducted to determine the acute toxicological profile of a compound. Using these data, lower dose subchronic studies are performed which identify the toxic endpoints of the drug. They will also determine the "no toxic effect" dose of drug, which will be used to test efficacy in other animal studies. Implicit in this movement from acute to subchronic doses is the assumption that a lower dose is safer. During this phase, initial studies are also conducted to find out how the body handles and excretes the drug (pharmacokinetics). Scientists then extrapolate from this animal data to the design of human trials.

As the drug continues to proceed through this maze of testing, other assays designed to assess teratogenicity and carcinoginicity (ability of a chemical to cause cancer) are conducted. The drug will then go forward into the earliest human clinical trials if the doses that prove effective in combating the disease are below those which produce adverse effects. For safety reasons, these early human trials are performed to validate the animal data and are not used to prove efficacy. *Implicit to this process is the complete acceptance of the dose-response paradigm where we acknowledge the fact that high or long-term drug therapy may be harmful, but low-dose or short-term therapy may be beneficial.* I am amazed that we cannot carry this time-tested logic to our foods and quit worrying about miniscule levels of the few farm chemicals that we occasionally find in some of our fruits and vegetables.

I should also add a caveat to this statement to assure the reader that I do not favor the presence of chemicals for the sake of chemicals. The chemicals we are talking about are used in food production for specific beneficial purposes. It is these low-level residues that persist in some foods, below federal limits defined as safe, that I am not concerned about. If there is no need for a chemical to be present in food, then obviously no one should put it there, and efforts should be made to minimize exposure in the most cost-beneficial manner. Finally, we are talking about chemical effects on food safety and *not* environmental safety, which is an entirely different issue.

In the drug approval process, after a safe human dose is established, larger scale human trials are conducted which pit the new drug against control groups receiving other therapy. These studies prove efficacy and safety in a larger and more diverse population of people. After the results of these trials are analyzed, the final step before approval by the FDA is large-scale multicenter trials (centers located throughout the country) that expose a very large population to the drug. A clear and statistically significant benefit must be proven to allow the drug to be marketed.

These studies, in addition to developing an entire arsenal of safe and effective drugs, have also validated the dose-response principle. In the early days of drug approval (prior to 1970), some inconsistencies were noted, and regulatory science has responded by expanding the gauntlet of tests required for approval. Examples include specific assays for immunotoxicity (damage to the immune system) and genetic toxicity (chemicals that may cause mutations). Some people have argued, and continue to argue, that this whole process is too restrictive. The cost to society could be a reduced number of effective drugs that might benefit a large number of people suffering from diseases for which there is no current treatment. Our society may be so risk aversive that many more innocent people suffer from a lack of drug availability than do those who actually experience adverse effects. Balancing this equation is next to impossible. We are increasingly being faced with a number of serious threats to public health. By being too restrictive and stopping drug development because of theoretical risks of adverse effects, we may face a generalized lack of availability of drugs when we really need them the most.

One development that has arisen out of this continually evolving science is the fact that the dose of drug given an animal is not even the critical measure. *It is the amount of drug that actually reaches the tissues that are affected by the drug that best correlates to predicting effect.* This is termed the *bioavailable* dose. In many cases a drug may enter the body but only a small amount is actually absorbed into the bloodstream. A large portion of this absorbed drug may then be destroyed by the liver or excreted by the kidney before it has a chance to exert its effects. The trick to getting a drug to work is to get a significant amount in fast enough to have an effect on the targeted tissue. That is why so many drugs can only be given by injection. Modern science is not smart enough to get large amounts of a drug into the body to have an effect by any simpler and noninvasive means. We will revisit this topic again when we discuss peptide drugs such as the infamous BST in milk as well as when we discuss the concept that *exposure* to a chemical does not equate to absorption and effect.

It is interesting to look at the results of some of these animal trials. Some drugs commonly have outstanding activity in isolated *in vitro* assays (tests done outside of the living organism; see full explanation later in the chapter), only to fail miserably when put in an animal because no intact drug makes it to the site of action. There may be a number of reasons for this. The first is that our body has been designed over the eons of evolution to protect itself from foreign chemicals. If we eat chemicals, the first step in its sojourn through the body is immersion in the "acid vat" we lovingly call our stomach. Many drugs never make it any further. If they do, they are absorbed into the body and immediately are delivered to the liver, our body's equivalent of a "chemical waste treatment plant." The waste, actually in the form of the chemical bound to a natural molecule such as an amino acid, is now more water soluble and is easily excreted by the kidneys into the urine without traveling any further in the body.

There is another principle that should be discussed when considering absorption of the chemical into the body—movement of the absorbed chemical from the bloodstream into the tissue. Like absorption, this movement is governed by the law of diffusion, which states that chemicals move from a region of high concentration to one of low concentration. Thus to get a drug to go inside the body, there must be a higher concentration on the outside (gradient or ramp) to "drive" the chemical in. Similarly, once the chemical is in the blood, it must build up to the point that it can be driven into the tissue. For some chemicals, "active transport" (a system is present to pump the chemical in) may occur; however, diffusion will always be a component of getting the chemical to the site of transport (the pump). This is a fundamental principle of science that applies equally well to environmental chemicals, where low-level exposure occurs. Chemical levels are often too low to drive the diffusion process. In contrast, occupational exposure to chemicals and pesticides may cause harm to the worker who is not properly protected because, in these cases, there is sufficient chemical present to "force" itself into the body down its concentration gradient.

This system of protecting our bodies from external chemicals works well at most doses encountered by people. The problem is with higher doses that overwhelm these defensive mechanisms. When this occurs, drugs now get to areas of the body that they normally would not reach, and often toxicity ensues. They may get there in a form that has not been detoxified by the liver. These effects often occur only at high doses and are a main reason why thresholds are present with exposure to toxic chemicals. However, the body may adapt to even these situations by stimulating more enzymes in the liver to detoxify the added burden. For most chemicals encountered, this is an additional protective effect.

A good illustration of this phenomenon is seen with another common toxicological exercise too often practiced by the American public: alcohol consumption. Most people appreciate that alcohol follows a dose-response relationship. Consume one beer, and nothing happens. Consume two and, to most people and even law enforcement agencies, there is no detectable effect or violative amounts appearing in your breath (an excretory route for alcohol). In fact, this low level of consumption has recently been shown to have beneficial effects on protecting against heart disease. However, as you consume more, especially if you do it rapidly, you overwhelm the body's defensive mechanism and neurological signs commence. This is commonly referred to as being "drunk." Again, this is practical toxicology and a clear real-world application of the dose-response relationship and the concept of thresholds.

Alcohol consumption can also be used to illustrate the dangers of directly extrapolating high-dose animal studies to humans. First, forget the fact that we know, even using bad statistics, that high-dose or chronic alcohol consumption is injurious to our health. We all know that, ultimately, our liver will be "pickled" and end-stage liver failure will occur. This was recently driven home to the public when baseball's all-time great Mickey Mantle received a liver transplant to replace his own liver, which received a lifetime of abuse. As the liver is being destroyed, its ability to detoxify absorbed chemicals and drugs is altered, and thus toxicity is often potentiated after exposure to normally safe levels of drugs and chemicals. Forget about this known effect. A major toxicological endpoint in laboratory studies of ethanol is that it is a known teratogen, a finding that has unfortunately been confirmed in humans.

If we take the laws and regulations through which we control pesticides and chemical additives in our food and apply it to the ethanol data, one could calculate the "safe" and "allowable" level of alcohol consumption that would protect us from birth defects or cancer. Our high-dose laboratory animal data already predict the effect of high-dose human consumption as animals also become intoxicated. Of course, this assumes that prohibition could be politically reinstated on the basis of teratology or carcinogenicity data rather than liver damage, deaths due to alcohol-related automobile accidents, or moral fervor. But it is teratology and carcinogenicity data that dictates our policy on pesticides. Dr. Bruce Ames has done such a calculation, and based upon the identical risk assessment process used to establish tolerance to dioxin (a potentially nasty chemical to be fully discussed later in this book), he has determined that only 1 beer every 8,000 years could be consumed without fear of causing birth defects. If you only drank 1 beer every 294 years, you would not have to worry about increasing your risk to cancer. When these calculations are done for all synthetic and natural chemicals in food, the relative risk of a glass of wine or bottle of beer rank at the top, followed by a host of natural carcinogens found in food, bacon, coffee, and finally at the bottom of this list (at some *million-fold safer level of risk*) are pesticides which are the main subject of this book. This level of anxiety associated with pesticide exposure is misplaced and takes our attention away from far more serious health concerns.

Before we completely leave alcohol, it is important to revisit the statement above that chronic alcohol consumption often results in chronic liver disease. Such livers are often incapable of metabolizing drugs and chemicals that would normally be eliminated from the body rapidly. Because of this damage to the liver, chemicals and drugs may now accumulate in the body and cause adverse effects due to the inability of the diseased liver to detoxify them. As we will see, when large-scale human trials are conducted to associate chemical exposure to toxicity, it is often only individuals with alcohol-induced liver damage that show adverse effects to chemical exposure. The alcohol-induced toxicity to the liver potentiates the adverse effects of chemicals metabolized by it.

We have a double standard and are legally selling and marketing chemicals with known adverse effects that, along with most of our natural food, have a risk of cancer millions of time higher than all the so-called chemicals polluting our food supply. We spend billions of dollars promoting the former and billions more protecting ourselves against the latter. If this is true, then why are we not all dropping dead like flies? The reason is that the dose is too low to have any effect! Low-level alcohol consumption presents a minimal risk, and some studies have even shown a beneficial effect. This is a clear-cut case where using toxicology improperly and ignoring the dose-response relationship leads to wrong predictions that are not borne out in human studies.

One class of compounds does deserve special attention—those chemicals that are very fat soluble. These include the older chlorinated pesticides, such as DDT, and some environmental contaminants, such as the PCBs (the polychlorinated biphenyls) and dioxins. In these cases, low levels of intact chemicals, if absorbed, may encounter the body's final defense mechanism. This defense is to "store" the compounds in fat and milk, a paradoxical strategy that the body uses to sequester away chemicals into a storage site (or depot) where they can do no harm to the rest of the body. It's conceptually the "prison system" of the body. The absorption and distribution into fat greatly reduces the concentration of chemicals at other body sites and thus serves to blunt the impact of the exposure. Similarly, this simple method of diluting the absorbed chemical often keeps the concentration below an effect level and gives the overworked liver time to try to destroy them.

However, the widespread presence of these synthetic chemicals in our body was alarming to many. Because of these concerns, most have been banned for a number of years, some even for decades. (Please consult the Appendix for a further discussion of persistence and the chemicals involved.) Therefore, the problem of persistent pesticides is an old one that has largely been solved! The amounts of these chemicals in our bodies and in the environment are stable or decreasing, and they should not enter into a discussion of the safety of the new generations of chemicals being produced.

Not to be remiss, there are other chemicals that actually become more toxic when metabolized by the liver. These *exceptions* to the rule will be dealt with later in the book when specific chemicals are covered. There are also documented interactions between chemicals, such that sometimes exposure to multiple chemicals produces enhanced effects. The classic example is the potentiating (enhancing) effect of tobacco smoking (in a dose-response manner) on almost all types of cancer. Eliminating this factor alone would tremendously reduce the incidence of cancer in this country. Another common confounding factor is chronic alcohol consumption discussed earlier that potentiates the toxicity of many chemicals through its effects on changing the enzymatic profile of the liver. However, for every additive interaction, there is also a protective or inactivating one where the effects of chemicals cancel out. This is the surprising conclusion of many of the chemical mixture studies conducted in the author's laboratory.

Let us assume we have a chemical that has managed to be absorbed into the body across the protective barriers of the skin or gastrointestinal tract and has either survived the destructive capabilities of the liver (or was inadvertently activated by it), or left the safe fat depots and actually reached a target tissue. These first defenses prevent a chemical from getting to the site of action. Alternatively, suppose you let your guard down once in your lifetime and drink one beer. Why have you not died? The main reason is that the body has a host of additional defense mechanisms that "kick into gear" to protect us from chemical assault by those molecules that have managed to make it to the site of chemical action. We possess these because of an evolutionary adaptation to living in a world where we constantly consume natural chemical carcinogens.

If the chemical is absorbed and bypasses our defenses, the chemical must then get into a cell or activate a receptor (a chemical switch) to have an effect. The binding to a receptor, which requires some specific chemical attribute, can then alter the cell's function or even activate a so-called oncogene that can result in cancer. Binding to a receptor itself is a dose-response phenomenon, and there is a threshold below which no effect is caused.

This is an important concept to remember. There are many natural and endogenous (coming from our own body) compounds that may bind to receptors but at levels too low to elicit an effect. *Although they may be present at the site of activity, they still do not produce an effect.* This is the crux of the debate on so-called environmental estrogens where certain environmental chemicals in high-dose studies are capable of binding to the female hormone estrogen receptors. However, at low doses, the chemicals cannot compete with the natural estrogens—such as estradiol—which are always present, even in males. The receptor's affinity for the natural hormone is much greater than for the synthetic chemicals. For wildlife, it is only in cases of massive environmental spills, high-dose laboratory animal studies, or exposure to therapeutic estrogen drugs, such as DES (diethlystilbesterol), where adverse effects in the intact animal or human may be expected. Again, it is wrong to use this data to extrapolate to low-dose exposure scenarios.

Even if the chemical activates the receptor, there are a number of additional cellular protective mechanisms that occur. They are related to so-called signal transduction—how the switch is wired—that are far beyond the scope of this book. Numerous antioxidants, which are found in our fruits and vegetables, and other scavenger chemicals are also available to bind up active chemicals or their metabolites before they can bind to receptors or do damage to the genetic material of the cell. One of these, the hormone melatonin, has recently become one of the latest health crazes purported to prolong our lives. Remember, many of these additional defensive systems are "standard equipment" in humans.

Other defense mechanisms are also available. Cells have effective transport systems which pump out foreign chemicals. Our cells are very good at repairing chemically damaged DNA (deoxyribonucleic acid; see Chapter 9 for a more detailed discussion of DNA) and even this event is not fatal. Only when the capacity for the cell to repair itself is overwhelmed does disease result. Again, dose is the determining factor. Finally, if all else fails, cells that become carcinogenic often induce an immune response to themselves and are eliminated by this route.When these mechanisms have been exhausted, or, in the case of extreme age, when they have become "worn out," disease occurs.

Most of these mechanisms of toxicity have been defined only at high doses in animal or *in vitro* cell culture studies where the protective measures have been inactivated or removed. The term *in vitro* refers to *studies that are conducted outside of the living body in an artificial environment.* These include such techniques as receptor assays, cell culture studies, bacterial mutagenicity tests, and other novel techniques of biotechnology. Recent developments have included hybrid receptor and DNA gene "chips" which link biological molecules to silicone chips that can dramatically expand the ability to detect which chemicals bind to biologically important receptors. Receptors have been described as the biochemical switches that activate responses. The chemical binding to a receptor "flips" the switch. In such *in vitro* systems, there is no problem getting high doses of drugs delivered to the cells since the body's protection is gone and one can literally pour the drug into the vial containing the biological targets. The doses used are often astronomical compared to exposure levels seen in a living person. However, when the results of these very sensitive techniques are used in risk assessment, the principles of exposure and bioavailability are often ignored. This occurs despite warnings from the assay's developers against their use in the regulatory arena. *In vitro* studies are appropriate if the proper concentrations are used or when the risk assessment takes into account exposure, bioavailability, and other factors reducing delivery of chemical to the active site where the receptors are located. Recent advances in genetic engineering will make them even more powerful. However, these assays in a test tube were never meant to serve alone as models for risk assessment because the effects only occur at high doses or under very artificial exposure conditions.

The problem goes beyond the act of taking *in vitro* data to predict an *in vivo* response—that is, a response in a living organism. Often in intact animal or human trials, the ridiculously high doses of chemicals used induce specific defense mechanisms that may do more harm than good against these unnatural high-dose exposures. For example, chronic irritation of continuous high-dose chemical treatment is often responsible for the cancer. If you were exposed to such irritating doses, you would leave the area. Our laboratory animals do not have that option. The stress of extensive handling has even been implicated as a contributing factor to the disease seen. Some studies are conducted in highly inbred strains of mice that are unusually sensitive or more susceptible to specific classes of chemicals because their receptors are more accessible or are present in increased numbers. Even control animals in these studies have much higher incidences of tumors than other strains of mice. There is no argument that these studies are important because they do shed light on *the mechanism of action* of these chemicals. Their use in risk assessment is the focus of the debate.

There is one facet of low-dose exposure that violates the dose-response paradigm—allergic reactions or hypersensitivity to minute quantities of chemicals. Most known allergens are natural in origin, and many of us have had experience with them. There are a number of well-documented drug allergies with penicillin being the most common. It has been estimated that 25% of the population has some type of allergy. Only a low percentage of individuals are allergic to a specific drug or to other natural compounds. The regulatory philosophy has always been not to withhold drugs because of this effect. There is nothing fundamentally different with pesticides. Our scientists will never develop drugs and chemicals that are absolutely safe in all 280 million inhabitants of the United States. There will always be a fraction of individuals who are allergic to something in the environment. Although this is unfortunate, much like other diseases, it is not completely avoidable. This phenomenon of extreme chemical sensitivity will be addressed when the so-called multiple chemical sensitivity (MCS) syndrome is discussed later in the book. There comes a point when the huge investment required to remove the offending molecule is too great and may inadvertently cause other deaths in the population, for example, by reducing availability of fresh fruits and vegetables. The better approach would be to further understand the mechanism of this effect in the affected individuals and design strategies to block their sensitivities much as allergists have been doing for decades to protect their patients from natural allergens.

Part of this concern over synthetic chemicals has arisen because of the dramatic increase in our ability to actually detect chemicals in our body and environment through advances in analytical chemistry, immunology and biotechnology. However, the mere presence of a chemical does not mean that an effect will occur or has already occurred. And are we ever getting good at detecting chemicals!

There have been numerous advances made in traditional analytical chemistry which have pushed the level of detection to parts per billion and even parts per trillion. The sensitivity is phenomenal, but what does it mean? Similarly, tremendous advances in immunology have harnessed the power of antibodies to detect molecules of drugs in biological or environmental samples. Biotechnology has cloned specific receptors to develop supersensitive chemical detectors. Some amazing advances have even coupled biological receptors with computer chips for chemical detection. Thus chemicals can be detected at levels millions of time lower than ever could produce a biological effect. We already know from laboratory animal studies and our discussion of statistics that much higher amounts of chemicals produce no effects. Since we could never prove a total absence of activity, and since the one in a million responder would be lost in the background noise of spontaneous disease, how should we react when chemists continue to find ever-decreasing amounts of chemicals in our food? A trivial and almost absurd analogy would be this: You know that if a one-ton brick fell on you, you would die. If you could detect a brick one millionth in size, or 0.032 ounces (just under 1 gram), would you be concerned?

This area of residue safety is a field in which I have had experiences since our laboratory has served as a national information resource for drug and chemical residues in food-producing animals. This service, the Food Animal Residue Avoidance Databank (FARAD), has published four technical compendia and maintains a national "hotline" for residue avoidance questions. For the last decade we have been compiling drug, chemical, and pesticide concentration data in meat and tissues to assure veterinarians and farmers that the animals they slaughter and the milk they produce are not contaminated with even minuscule amounts of drugs or chemicals. We are protecting against the 0.032 ounce bricks. Unlike drug treatment in humans, when a veterinarian treats an animal, the dose must be selected that cures the disease, does not produce toxicity, and does not produce residues in the milk or meat. We eat our patients and thus must eliminate any residual drug to prevent human exposure. Your physician treating you does not worry that leftover drug remains in your body!

Much of this regulatory zeal was driven by the Delaney Clause, which amended the Food, Drug and Cosmetic Act of 1938. This clause forbids the addition of any amount of animal carcinogen to the food supply. This was originally based on our belief at the time that even one molecule of a carcinogen could cause cancer in humans. This concept was largely influenced by theories of radiation-induced cancer. Thresholds were not allowed. As we discussed, this is no longer considered valid since the processes of absorption, distribution, elimination, metabolism and cellular defense, and repair mechanisms make this possibility far less than remote. However, the United States

Congress mandated that there be zero tolerance for any chemical identified in any animal test as being a carcinogen at any dose. In the late 1950s analytical chemistry was such that parts-per-million detection of chemicals was considered an extremely sensitive assay. We can now detect a million times less chemical, and at levels that were never anticipated when these laws were passed. Using our brick analogy, we are now detecting bricks weighing one millionth of a gram (a nanogram). Our brick is now microscopic, and even the fictional inhabitants of Lilliput, from *Gulliver's Travels,* would not fret!

Congress and the regulatory community realized that analytical chemistry was getting ahead of the biology, and in a 1968 amendment to the Delaney Clause—the DES Proviso—they linked the allowable limit to the analytical method. Thus carcinogenic drugs could be used in animals if one could not find residues in the tissues. Beginning in the early 1970s, the FDA began working on a sensitivity-of-method approach to regulating residues, which was ultimately published in 1987. This approach, currently in use, defines a threshold residue level (see Appendix A for how these are set), below which food is deemed "safe" even if analytical methods can detect a residue. The residue level is set to a risk of cancer of 1 in 1,000,000. The scientific debate, which lasted over a decade, is over, and almost everyone accepts that detection below this tolerance adds an immeasurable level of risk to the food supply. Analytical detection was finally uncoupled from a potential to cause biological effect. Yet even today, it is detection at these Lilliputian levels that causes the media hysteria and a tremendous amount of unfounded anxiety in the public's psyche.

This "no-tolerance" philosophy has removed a great number of chemicals from our food supply and has probably forced the development of even safer and more effective therapeutics. All drugs approved for animals today must undergo extensive residue trials, and safe tolerances, having up to 1000-fold safety factors, are established. Information programs such as FARAD, coupled with testing programs, ensure that food products have tissue residues *below* these safe levels. When surveys are done which indicate that a drug has been detected in our food, there should not be panic. *Detection does not equal toxicity.*

A practical observation is that the drug levels detected in the edible tissues and in the milk in food-producing animals produce no biological effects to the animal itself! This even includes animals that are treated for long periods of their life, such as dairy cows. However, some alarmists would have one believe that if people ate meat from these same animals, some adverse effect might occur to them. I can assure you that I do not lose sleep over this risk. I would limit my consumption of meat and milk because of its fat and cholesterol content, not because of any drug or chemical residues. Bacterial contamination of certain meat products is a much more substantial and real threat to our safety. Public concern and research should be focused on this aspect of food safety and not on theoretical risks of minuscule levels of chemicals.

This research, and the practice of food-animal medicine, has generated a database that clearly shows just how distinct the effects of drugs are at toxicological levels, at lower therapeutic levels, and at extremely low but detectable residue levels. Each amount of drug occupies different regions of the dose-response hierarchy. There is very little overlap between the ranges. This hierarchical behavior of chemicals at different concentrations is analogous to some fundamental principles on the order of nature, promulgated by the Nobel laureate physicist Dr. Phillip W. Anderson in 1972. He proposes that reality has a hierarchical structure with each level independent, to some degree, of the levels above and below. He writes, "At each stage, entirely new laws, concepts, and generalizations are necessary, requiring inspiration and creativity to just as great a degree as in the previous one." We would be well advised to apply this insight into the problem of low-level chemical action. There is a barrier which prevents extrapolation of data from artificial high-dose non-physiological experiments (the realm of toxicology) to the physiological level where normal homeostatic or life-maintaining mechanisms (the realm of pharmacology) are operative, and to the level where molecular interactions (the realm of low-level exposure and residues) that have little biological consequence may occur.

We must realize that the push to eliminate low-level chemical contamination at all cost could have significant, unforeseen adverse consequences. This might clearly be illustrated in veterinary food-animal medicine where the drive of FDA regulations over the last two decades was to reduce the amount of drugs used "at all cost" in food-producing animals so as to minimize the drug load in food products and thus minimize human exposure. Drug approval strategies were aimed at identifying the lowest possible dose that would treat an animal so that minimum residues were produced. I participated in a recent April, 1995, FDA-sponsored workshop designed to develop new paradigms of drug approval. The realization surfaced among many participants that this policy also resulted in the use of less-than-optimal antibiotic therapy to treat food-producing animals because of the fear of residues. This may have been one factor that contributed to the emergence of antibiotic-resistant pathogens in veterinary medicine. Rational or prudent therapy, based on sound scientific principles and medicine, is required, not minimalist therapy driven by theoretical risks. Were public health and animal welfare served by these actions?

The scientific establishment has largely put to rest the single-molecule, no-threshold theory of cancer. Let us do the same! Regulatory agencies, such as the EPA, are now putting more science in risk assessment. Regulations promulgated in 1996 accept specific levels of risk as being negligible and lower levels as being beyond our ability to quantify. Let us hope that the public does the same and leaves the theories of the past in the past. Let us learn from the mistakes of the past but live in the present and plan for the future.

As Paracelsus indicated in the sixteenth century, all chemicals may be poisons; it is the dose that makes a difference. I believe that it is a dangerous exercise to blindly extrapolate high-dose animal toxicology data to low-dose residual exposures. As will be discussed later, approaches which factor in mechanism of action, detailed exposure information, pharmacokinetics, and metabolism data offer a more rational approach to risk assessment. Finally, our exploding growth in analytical chemistry and our prowess in immunology have long ago surpassed the ability of biology and medicine to interpret their significance since the biological effect level is well within that of spontaneous disease. Detection of a chemical in food does not imply risk.

> The thing that numbs the heart is this,
>   That men cannot devise
> Some scheme of life to banish fear
>   That lurks in most men's eyes.
>
>   (James Norman Hall, "Fear")

The Pesticide Threat

The reader now has some of the basic concepts necessary to objectively assess the real risk of consuming fruits and vegetables contaminated with residue levels of pesticides. There are a few more tools needed before we can solve the puzzle. Prior to addressing the true "toxicology" of these chemicals, we should become familiar with the actual culprits.

What are pesticides? The *Merriam Webster's Collegiate Dictionary* (1994) defines a pesticide as "an agent used to destroy pests."

An agent is a chemically, physically, or biologically active principle. We will use the word agent as being synonymous with a chemical or drug throughout this book. We are not concerned with physical techniques or the use of predatory insects or bacteria to kill pests. We are focused on chemical hazards. But then, what is the definition of a pest? Again, according to *Webster's*, a pest is defined as:

1. An epidemic disease associated with high mortality: spec. Plague
2. Something resembling a pest in destructiveness; esp. a plant or animal detrimental to humans or human concerns (as agriculture or livestock production)
3. One that pesters or annoys: Nuisance

For the purposes of this book, we will combine these and define a pesticide as *a chemical agent used to destroy a plant or animal detrimental to agriculture or livestock production.* I will not cover agents used to treat infectious diseases since I would classify those as therapeutics. They would include antibiotics or antimicrobial drugs, antivirals, and antifungal drugs. Some workers in the field would argue that these drugs should be included; however, they would add little to the debate. The question of antibiotic residues is another ball-game that would take an additional book to discuss due to the potential they may have of modulating bacterial resistance. The only relevant concern for the purposes of our discussions is when these compounds appear in the meat or milk of food-producing animals where they might cause residues and thus require a public health assessment.

The focus of this book is chemical pesticides that occur primarily on plant-derived foods such as fruits and vegetables. These are the compounds with the greatest public awareness. Part of this attention could be related to what happens when one reads the dictionary and notices that "plague" is associated with these terms! These are serious diseases. As a side note, the plague that killed a large percentage of the population of Europe during the Middle Ages was carried by fleas. This would not be such a threat today because of the existence of pesticides.

Thus we focus on chemicals used to treat plant and animal pests that are detrimental to agriculture and food destined for human consumption. The term *pesticide* is an umbrella for numerous compounds used to eliminate many types of pests. The bulk of the toxicology literature is concerned with the following:

- Insecticides—pesticides that kill insects
- Herbicides—pesticides that kill unwanted weeds
- Fungicides—pesticides that eliminate fungi and mold, which themselves produce lethal toxins if allowed to contaminate our food
- Molluscicides—pesticides that kill slugs and snails
- Rodenticides—pesticides that kill rodent pests such as rats and mice

Insecticides are chemically classified into a few major groups—the chlorinated hydrocarbons, organophosphates, carbamates, and pyrethroids. The vast majority of attention relative to food safety will be focused on insecticides, herbicides, and fungicides which have become integral components of modern high-yield agriculture.

When did the pesticide scare begin? Unlike other questions posed in this book, this one has a single answer. It began with the publication of Rachel Carson's *Silent Spring* in 1962. This seminal book has been credited with launching the environmental movement that ultimately resulted in the establishment of the Environmental Protection Agency (EPA). It has also been the rallying point for much of the "residual" fear of pesticides that has persisted for four decades.

I remember reading this book as a "classic" when I was an undergraduate in the early seventies, and, like most of my contemporaries, I was moved by its arguments. I became a toxicologist and essentially devoted my profession-al career to understanding how chemicals interact with humans and animals. The book was typical of the earth-friendly movements of the sixties, which reacted against the coldness and efficiency of the military-industrial complex. *Silent Spring* is an emotional plea for reducing chemical usage because of what was perceived by Carson as a poisoning of our environment with *persistent* chemicals. The book's title is based on a scenario described in the opening chapter titled "Fable for Tomorrow" that depicts a world ravaged by synthetic chemicals, a world devoid of birds and thus characterized by a *silent spring*. I quote her description of an affected town:

> There was a strange stillness. The birds, for example—where had they gone? Many people spoke of them, puzzled and disturbed. The feeding stations in the backyards were deserted. The few birds seen anywhere were moribund; they trembled violently and could not fly. It was a spring without voices. On the mornings that had once throbbed with the dawn chorus of robins, catbirds, doves, jays, wrens, and scores of other bird voices there was no sound; only silence lay over the fields and woods and marsh.

These are eloquent and powerful words. However, it describes a vision of a future resulting from unabated use of pesticides. Rachel Carson was a marine biologist working for the U.S. Fish and Wildlife Service and an author of two other successful books—*The Sea Around Us* and *The Edge of the Sea.* What troubled her and compelled her to write *Silent Spring* was her belief, as the first true ecologist, that humans evolved in an environment that did not include synthetic organic chemicals—the "elixirs of death," as she referred to them. She believed that it was a mistake to pollute the earth with these poisons, since humans and animals did not have the biological defenses to cope.

She was concerned primarily with the infamous persistent chlorinated hydrocarbon DDT {1,1,1-trichloro-2,2-bis (p-chlorophenyl) ethane}. During the sixties, DDT was being used to combat Dutch Elm disease by targeting the bark beetles which were the responsible culprit. However, at the time, there was evidence that DDT also killed many other insects and bioaccumulated in the earthworm. Robins and their young would then eat earthworms and suf-fer the agonizing signs of acute chlorinated hydrocarbon poisoning. This was generally followed by death. She reported similar scenarios of bird kills across the country. The demise of the peregrine falcon was blamed on DDT. Sampling of foods demonstrated DDT residues, and it was discovered that high concentrations were found in animals and even in human adipose tissue and milk (see Chapter 2). Her conclusion was that the earth was being poi-soned with these "deadly chemicals," and that it must stop.

I will not debate the validity of the arguments presented because it is easy, some 40 years later with what amounts to almost infinite hindsight, to realize that many of her concepts were based on insufficient data and knowledge of toxicological mechanisms. Others have taken on this task, even up to the present, and the list of references provides ample reading material (see Avery, Bast et al., and the classic by Efron for a sampling). *However, this is not the purpose of my book!* I believe that there is sufficient evidence that our health is improving and the fears presented in *Silent Spring* never materialized. This may in part be due to the increased awareness that she brought to the inherent dangers of persistent organic chemicals. However, the chemicals we deal with today are significantly different than those used in Rachel Carson's time, and the field of toxicology, which essentially did not exist then, now is a powerful force to avoid making similar mistakes in the future.

Many of the assumptions under which she was working have now been refuted. Her mechanism of action for the adverse effects of pesticides was that continuous exposure slowly damaged our cell's metabolic machinery. This is fundamentally wrong and has no serious scientific proponents today. Curiously, similar arguments and even *Silent Spring* itself are invoked today by people suffering from Multiple Chemical Sensitivity (MCS) syndrome (this topic to be dealt with in more detail later in this book).

Similarly, some of the previously mentioned authors have written that her ecological arguments were unfounded. Bird populations were actually *increasing* during this period. For those that did decline—like robins—the cause may have been due to eating seeds treated with the pesticide and fungicide methyl mercury! They also argue that many of the ecological disasters were due to habitat loss from human population incursions secondary to development. Peregrine falcon numbers were well on the decline *before* DDT was introduced. Some of these workers quote studies that DDT did *not* result in thin egg shells. This phenomenon could also have been caused from the environmental pollutant PCB, a concern presently echoing in the debate on the safety of environmental estrogens. (Note that these compounds are no longer of concern since they have long ago been removed from production.)

I believe that it is irrelevant whether the arguments about DDT in the sixties were true or not. Prior to this time and her writings, we did not have a concept of ecology and probably would have suffered some environmental disaster. I strongly believe that Rachel Carson caused an awakening that had many positive effects. Her writings definitely influenced me and helped determine my present career path. However, these are issues of the past and progress has been made. Some have argued that even if DDT did not kill the robins, methyl mercury probably did, and this is an ecological concern. Similarly, the common pesticides in use in the forties contained arsenic, mercury, fluorine, and lead. These were dangerous chemicals, and replacement by even the chlorinated hydrocarbons would be considered an improvement. A similar argument holds today when attacks on the safety of biotechnology must be countered with the continued use of chemical pesticides which would occur if chemicals were not replaced with the products of this new industry.

This is a good point to mention that the only commonly used pesticide ever *proven* to be a human carcinogen was lead arsenate. DDT was postulated to be a carcinogen on the basis of weak-to-nonexistent evidence. This myth has been firmly implanted in the American consciousness. These concerns resulted in developing rigid standards of good laboratory practice that helps ensure that the safety of chemicals are appropriately tested and the integrity of research can be guaranteed. *We have made significant progress since the sixties. We cannot invoke the "Spirit of Silent Spring" for every new chemical we encounter!* Doing so will have its costs because the benefits of pesticide use will be forsaken.

Synthetic chemicals do not have a monopoly of being able to cause ill effects to man and animals. There are a host of natural carcinogens, such as alflatoxins, to which we are constantly exposed. Similarly, many of the synthetics were designed specifically to mimic natural plant and animal toxins that have killed numerous people throughout history. At high doses some of these compounds are poisonous to human and other nontarget organisms (see Appendix A for individual chemical details; also see the previous chapter for the limitations of extrapolating from high-dose toxicology to low-dose residue concerns).

Insecticides were developed to kill insect pests that are themselves proven to be hazardous to human health. Pesticides have generally been useful to the human race. Insecticides have forever banished the threat of locust plagues that swept nineteenth century America. Also, they have been a primary reason why American agriculture is dominant in the world, and they allow us to debate what to do with our grain surpluses rather than argue over how to increase agricultural productivity to feed our people. Significant occupational exposures to these compounds continue to exist around the world, and this exposure scenario may result in toxicity when inadequate safety precautions are applied.

There was a tremendous cost to removing DDT from the world markets. I believe that the writings of Dixie Lee Ray are the most poignant, regarding the cost to society of eliminating this single insecticide. DDT worked, and its persistence after application decreased the need for continuous exposure. Its use in World War II saved countless Allied soldiers from dying of typhoid fever since its use on the soldiers killed the lice that were the vectors of this historical scourge of humanity. Ray estimates that more soldiers died of typhus in World War I than from bullets! Similarly, it is widely documented that before the advent of DDT, 200 million people worldwide were annually stricken with mosquito-carried malaria. About 2 million died each year. After the advent and use of DDT, these numbers dramatically dropped.

In books written by both Dixie Lee Ray and Dennis Avery, a chilling example of the effect on human health of DDT's withdrawal is presented. In Sri Lanka (formerly Ceylon), there were 2.8 million cases of malaria prior to the advent of DDT and only 17 by 1963 when its use was widespread. Because of the rising anxiety in the United States about the theoretical possibility of DDT carcinogenesis, spraying was halted in the mid-sixties. In 1968, the number of malarial cases in Sri Lanka returned to 2.5 million. This is a tragedy that must not be repeated. Were any lives saved from DDT-induced cancer?

In fact, restricted use of persistent pesticides such as DDT has led to the resurgence of other lethal arthropod-transmitted diseases such as Leishmaniasis, whose incidence is rocketing at the dawn of the twenty-first century. Although we all would like to believe otherwise, modern society is still faced with a significant threat from insect-borne diseases. Just note the recent resurgence of once-rare diseases such as viral encephalitis and West Nile Virus in New York City. These diseases cannot easily be prevented by either vaccination or new antiviral drugs, for none exist. They can only be controlled by prudent use of synthetic chemical pesticides targeted against the insect vectors. Banning the use of pesticides for theoretical risks will only result in a real increase in death tolls from largely preventable tropical diseases. Additionally, due to the almost certainty that insect pests will develop resistance to any one specific class of insecticide, it is crucial that other pesticides remain available for future use.

Recently, these concerns have been extensively addressed by the international community. In 2000, a treaty on persistent organic pollutants signed by 122 countries will ban or phase-out 12 long-lived pesticides and other toxic chemicals. These include DDT, aldrin, dieldrin, endrin, chlordane, heptachlor, hexachlorobenzene, mirex, toxaphene, dioxins, furans, and PCBs. However, a reprieve is given to the use of DDT to control malaria in Africa, Latin America, and Asia for the reasons presented above concerning their overwhelming efficacy in controlling this deadly disease.

The other cost to all of us by removing DDT and similar insecticides is the fact that as a class of compounds, they had low acute toxicity to farmers and applicators using them. The compounds that replaced them, the organophosphates, did not persist in the environment but did result in many acute deaths, mostly due to occupational scenarios. They are just more toxic on an acute, short-term basis than are many other classes of insecticides. I personally work with these compounds from the perspective of studying dermal exposure and toxicity to the skin. Let me reiterate that at high concentrations, pesticides are potent chemicals. This is especially true if you spill them on your skin! However, this is a different issue than eating food contaminated with a trace level of a pesticide. But I am getting ahead of myself; I will deal with *real* toxicology in Chapter 7 and Appendix A, which specifically overviews the high-dose toxicology profiles of many pesticides.

We cannot turn back the clock and correct past mistakes. On the whole, it is probable that *Silent Spring* stirred an awareness of human ecology that would not have occurred otherwise. Similarly, this heightened awareness may have protected people from other more deadly disasters by prompting us to ask more probing questions about chronic toxicity and potential environmental impact as we introduced new chemicals. However, the cost in human suffering was also great. DDT will probably never return to the United States unless massive insect infestations require it. The environmental awareness and regulations that resulted are likely to stay with us. However, DDT must be a dead issue when debating risks of modern chemical exposure in food!

DDT is often indirectly used to heighten the public's fear of any chlorinated chemical. As I will discuss in the final chapter, it is imperative that we confront such chemophobia with facts. Decisions must be based on the most current science possible and common sense; otherwise, irrational fear may prevail. I do concede that DDT is persistent and toxic to our collective psyche because its memory is invoked as a battle cry for any chemical crusade. It has been invoked in the recent movement to ban chlorine from our drinking water as both DDT and organic complexes formed from the chlorination process are *chlorinated organic chemicals.* The goal of this movement was to protect us from being exposed to low levels of chlorinated chemicals that spontaneously form in our water supply. The argument goes that since they have chlorine in them, they might share the known hazards of DDT and other chlorinated chemicals such as dioxins and PCBs. The logic is that since DDT was banned, it must have been hazardous. The national Society of Toxicology is on record against this ban. These are the experts whose members include active environmentalists as well as academics and industrial toxicologists. There is no toxicological problem!

We cannot compound one mistake upon another and replace a theoretical-to-nonexistent risk of cancer with a real risk of waterborne bacterial epidemics. Removing chlorine from our water supply could result in a public health catastrophe with cholera outbreaks rivaling the greatest epidemics of past centuries. Some have estimated upward of 9 million deaths per year. We cannot allow pesticide phobia to kill people! This is the crux of my argument. We have serious health problems to confront as we live in the twenty-first century. We will need all of our tools,as well as the best science possible, to confront them. We have made significant progress in combating disease. Let's not sabotage this progress with unfounded hysteria.

The memory of DDT continues to be invoked in the nineties to support pesticide restrictions. This can be appreciated by turning on the television, reading newspapers and magazines, or perusing any one of a number of recent books (see Lawson or Setterberg and Shavelson as examples). In these books, the public clearly believes that we continue to be poisoned by terrible pesticides, with DDT being given as the classic example. It is long past the time to put this fear to rest.

The most succinct analogy to the fallacy and even danger of using 1950 and 1960 toxicology in 2002 is to apply this logic to modern drug safety. If one truly believes that all pesticides are dangerous and related to DDT, the next time illness strikes, one should shun modern medical treatment. The physician should be instructed to consult only medical texts from the sixties, and avoid prescribing any post-sixties drug because the safety and efficacy of all of our modern medications are based on the same chemistry and toxicology that has produced the new generations of pesticides. Let science move on and focus on the public health issues that do confront us in this new millennium. These present us with challenges enough!

Now that we know where the public concern for pesticides originated, we should assess whether pesticides are of concern in 2002. We will accomplish this by reviewing the nature of exposure for someone eating fruits and vegetables today. It is important to go through this exercise so that we put the continuing pesticide hysteria in context and assess whether past arguments are relevant to the present. Fortunately, there is good data to identify the pesticides to which we are exposed

By mandate of the U.S. Congress, the FDA is required to randomly sample food products in the United States as well as those imported from other countries. This survey covers 366 pesticides, an increase of 40 pesticides more than when this author first examined the data for 1992. These include currently marketed as well as older chemicals no longer used in the United States—for example, DDT and its metabolites. This program began in 1986 and is one component of the food safety system that monitors consumer exposure. The program is continually expanding, primarily in import areas where residues were detected in earlier years. It is an adaptive system designed to focus on and correct violations. We will concentrate on the latest reports available in 2001 that are most relevant to our present situation. Table 3.1 is such a compilation for 1999 using fresh fruit and vegetable consumption data compiled by the USDA and the results of the FDA Residue Monitoring Program for the same year. Data on total dietary exposure, which includes processed food, is also addressed. Note that these surveys were conducted by different agencies and thus some minor produce categories did not match across studies. For example, the United States Department of Agriculture (USDA) classifies cantaloupe and watermelon as vegetables while FDA classifies them as fruits.

It must be stressed that these are analytical studies to assess exposure and are not surveys of levels that are known to be toxic to humans. I believe it is important to define the scope of the problem and the types of chemicals that are present to form a factual basis upon which a discussion of real risk can be made.

The analytical methods used are state of the art and detect chemicals at concentrations well below the EPA's established tolerances, which range from 0.1 to 50 parts per million. Generally, residues at or greater than 0.01 parts per million (10 parts per billion) are detectable, and for some chemicals the sensitivity may even extend down to one part per billion! Residues are scored as follows:

- *Trace*—present above background but not quantifiable
- *Detectable*—quantifiable but lower than tolerance levels
- *Violative*—levels greater than the established tolerances or just detectable if no tolerances are established

A tolerance is a concentration that is deemed *safe* for *lifetime human consumption*. The reader should consult Chapter 6 where tolerances are more completely discussed. The data tabulated are from the most complete year available (1999) published in 2000. The reader is directed to the World Wide Web sites of the USDA (www.usda.gov) and FDA (www.fda.gov) for more current data as it becomes available.

It is important to recognize that levels exceeding the tolerance do not imply an adverse health effect. These surveys indicate that for United States produce in the last year of the twentieth century, violative levels when detected hover around 0.7 to 2.9% of all samples analyzed for each category of produce with the vast majority being zero. For produce imported into the United States, the violative levels when detected range from 1.0 to 4.8% within produce categories. Some 9,438 samples were analyzed to compile these data with 3,426 being from domestic produce grown in the United States. *Across all produce groups, no violative residues were found in 99.2% of all domestic samples and 96.9% of the imported samples.* Of the greater than 350 pesticides assayed, only 89 were detected. Pesticides could be detected at below-threshold levels in about one-third of all samples. In the Total Diet Study conducted by FDA on finished products selected to represent a typical consumer's consumption patterns, prepared food was analyzed for trace levels of some 200 pesticides at a sensitivity 5–10 times lower than the regulatory surveys above. Using this exaggerated technique, only 104 residues were found in some 1040 samples. The 5 most prevalent chemicals were DDT, methylchlorpyrifos, malathion, endosulfan, and dieldrin. This survey does illustrate the persistence of pesticide residues of DDT and dieldrin, which were banned from active use many years ago. Their source is most likely due to inclusion of animal products where chlorinated hydrocarbon compounds accumulate in fatty tissues (see Appendix for complete discussion).

These exposure levels to specific samples must then be compared with actual consumption of produce. This is estimated in table 3.1 by looking at per capita produce consumption published in the *2000 Statistical Abstract of the United States*. In 1998, the average American consumed 700 pounds of fruit and vegetables, approximately 60% of which were vegetables. The majority of this produce was grown domestically. This consumption includes produce directly consumed by individuals as well as that processed into finished products. Because of the nature of the American food production system, very few individuals are exposed to produce originating from the same sources. When these consumption figures are combined with the incidence of violative residues, just a minor percentage of produce would be expected to contain violative residues. The actual exposure of an individual would depend upon specific produce consumption patterns.

These numbers are averages across multiple vegetable and fruit categories. The majority of fruits and vegetables had zero violative residues with the average increased by specific offenders. The more important question is how did the fruits and vegetables that we consume the most fare in this analysis? If one closely examines Table 3.1 and looks at the major produce categories based on consumption (and thus real exposure), we would identify apples, bananas, cabbage, cantaloupe, carrots, celery, corn, cucumbers, grapes, lettuce, onions, oranges, peppers, potatoes, tomatoes, and watermelons. The vast majority of this produce had *no violative* residues detected. No residues of any kind were even *detected* in 60–70% of all samples analyzed. The major produce groups that account for the largest percentage of fruits and vegetables in our diet had rates of violative residues approaching 0%!

What is the state of our fruits and vegetables relative to the occurrence of residues from some 360 pesticides? Basically, anywhere from less than 1% to a maximum of 4% of samples contain a violative residue of some pesticide. Most of the fruits and vegetables with the highest human consumption have no violative residues. The pesticides detected are from numerous classes of some 90 diverse chemicals.

Of even more importance to the potential toxicological implications of these surveys, FDA has often targeted sampling of unregistered pesticides in imported products. These studies were driven by a fear that our shores were being invaded by produce contaminated with nonapproved chemicals that we were not monitoring. In a 1993 study, a number of specific unregistered pesticides on imported products were identified as being used based on surveys conducted in these countries. Residues were not detected on any of the imported fruits, vegetables, or fruit juices sampled. A similar analysis was performed for specific targeted unregistered pesticides—which would, therefore, have no tolerances on domestic produce and yielded no violative residues. In 1999, a study was done on imported grains from Canada, and only one sample

TABLE 3.1. Violative pesticide residues in selected fruits and vegetables (1999), compared to United States per capita utilization from domestic and imported sources in 1998.

	Per Capita Utilization (pounds)	Violative Residues Domestic (% samples)	Violative Residues Imported (% samples)
Fruits			
Apples	19.2	0.0	0.0
Bananas	28.6	0.0	0.0
Cantaloupe	11.3	0.0	0.0
Grapefruit	6.0	0.0	0.0
Grapes	7.3	0.0	1.0
Oranges	14.9	0.0	2.4
Peaches	4.8	0.0	0.0
Pears	3.4	0.0	0.0
Pineapples	2.8	0.0	1.4
Plums/Prunes	1.2	0.0	0.0
Strawberries	4.1	2.4	3.7
Tomatoes	17.4	0.7	1.6
Watermelon	14.5	0.0	3.0
Vegetables			
Asparagus	0.8	0.0	2.2
Beans	1.7	0.0	3.8
Broccoli	5.6	0.0	3.2
Cabbage	8.9	1.6	3.9
Carrots	13.6	2.9	0.0
Cauliflower	1.6	0.0	0.0
Celery	6.2	0.0	4.8
Corn	9.0	0.0	0.0
Cucumbers	6.7	0.0	0.8
Lettuce (Head)	22.8	0.0	0.0
Mushrooms	2.5	0.0	1.7
Onions	18.5	0.0	2.4
Peppers	6.4	2.6	2.5
Potatoes	47.8	2.6	0.0
Sweet Potatoes	4.1	0.0	0.0

% Violative tabulated as those over tolerance or detected if no tolerance established.
Produce categories do not exactly match in listings for residues versus utilization.

had detectable pesticide, which was well below U.S. tolerance. Additional surveys were conducted on pears and tomatoes showing lower violation rates than seen for earlier years. If one does not trust the FDA, various states with agricultural interests have maintained their own monitoring programs. During the same time period, only 1% of samples surveyed by these state programs contained violative residues.

It must again be stressed that violative means greater than a threshold established as safe by a regulatory agency. As will be discussed in Chapter 6, the threshold is based on an extrapolation and risk assessment process using high-dose animal studies with a 100 to 1,000 times safety factor built in. These thresholds also assume continuous consumption of the product in question with this level of pesticide always present on every piece of produce. With a 1% incidence in samples, this overestimates exposure 100-fold! This would also require that all produce come from the same source, as it would be extremely unlikely that contamination with the same pesticide would occur across different producers and distributors. Modern food production, processing, and distribution practices make consumption of any food product from a single source highly unlikely!

What these data do provide is a basis for determining what kinds of compounds are present in today's diet and roughly our expected exposure. These are incidences of pesticide detection and not surveys of pesticide-induced toxicity. This point was originally stressed in an introduction to an assessment of our analytical methodology for residue detection conducted by the watchdog agency of the U.S. Congress, the Office of Technology Assessment. This is the group that exists to make sure that the Executive Branch of the government, represented by agencies such as the FDA, EPA, and USDA, are doing their job. I quote:

> Although increased interest exists in improving analytical methods, no consensus has yet developed on the importance of doing so. In contrast to the general public's uneasiness over pesticide residues in food, the Federal agencies responsible for regulating foods do not have the same level of concern for the situation as it exists. Based on the low violation rates found in food under current testing programs, the Food and Drug Administration (FDA) of the U.S. Department of Health and Human Services and the Food Safety and Inspection Service (FSIS) of the U.S. Department of Agriculture consider that pesticide residues in food is not the most important food safety issue.
>
> (OTA–F-398, pg. 4, October, 1988).

Essentially, this report says that since we already can detect pesticides at levels well below any real biological risk, what is the point of pushing our limits even further? If what we detect now is too sensitive, what possible use will it be to detect even lower concentrations? I absolutely agree with this assessment made several years ago as it is even more relevant in 2002. However, as I discussed in the Introduction, some continue to sound an alarm well into this new century that we are being poisoned with pesticides. Recall the panic which occurred a few years ago when trace levels of Alar® (daminozide) were detected in apples and some by-products. In this case, Congress and the Executive Branch, as well as numerous panels of experts, agreed that there was no problem. Detection of trace levels of pesticides in baby food sounded a similar alarm while I was completing the first incarnation of this book. Why do these issues still persist? I agree that in our society the public has a right to know and that the press is obligated to report it, but there must be a rational filter applied and care taken not to induce hysteria since the results may be more dangerous than the cure.

A separate indication of exposure is the assessment of the actual usage of pesticides in agriculture in the 1990s. This will give us a different view on the kinds of chemicals being used and provide a basis for comparison to earlier times, like those of Rachel Carson, when usage patterns were different. This approach is necessary since random monitoring programs, such as the FDA project discussed above, were not instituted in earlier years and, therefore, no basis exists for comparisons across time. Analytical methods were different, and that alone would preclude a fair comparison. Table 3.2 is a tabulation of the top pesticides used in agriculture in 1974 versus 1993 as compiled by the EPA's Office of Pesticide Programs. Very similar to the situation above, these numbers reflect total agricultural usage and include some *nonfood* crops. In fact, 75% of all pesticides targeted against insect pests are used on *corn* and *cotton* crops. Note that corn is one vegetable with a 0% incidence of violative residues as tabulated in Table 3.1.

A close examination of this usage table shows that the herbicide atrazine has held the number one spot for the last 20 years. This compound is often used in nonfood crops such as cotton. However, some compounds are notably absent from the 1993 tabulation, including toxaphene, methyl-parathion, carbaryl, propachlor, DSMA, linuron, aldrin, carbofuran, chloramben, maneb, sodium chlorate, propanil, DBCP, malathion, dinoseb, and chlordane. It should also be noted that DDT, the pesticide on which *Silent Spring* was focused and which continues to be used as the example of man's folly with chemicals, is not on either list. The one generalization that can be made concerning this list of deletions is that most of these pesticides are more persistent than the compounds that replaced them and a significant number of them are chlorinated hydrocarbons. It is important to keep these compounds in mind since popular press accounts of pesticide contamination often quote incidences of exposure to unnamed pesticides and then focus most discussion

TABLE 3.2. Top 25 Pesticides Used in 1974 and 1993.

Rank	1974	1993
1	Atrazine	Atrazine
2	Toxaphene	Metolachlor
3	Methyl Parathion	Sulfur
4	Sulphur	Alachlor
5	Alachlor	Methyl-bromide
6	Petroleum Oil	Cyanazine
7	Dichloropropane/ene	Dichloropropane/ene
8	2,4-D	2,4-D
9	Butylate	Metam sodium
10	Trifluralin	Trifluralin
11	Carbaryl	Petroleum Oil
12	Propachlor	Pendimethalin
13	DSMA/MSMA	Glycophosphate
14	Parathion	EPTC
15	Linuron	Chlorpyrifos
16	Aldrin	Chlorothalonil
17	Carbofuran	Propanil
18	Chloramben	Dicamba
19	Maneb/Mancozeb	Terbufos
20	Sodium chlorate	Bentazone
21	Propanil	Mancozeb
22	DBCP	Copper hydroxide
23	Malathion	Parathion
24	Dinoseb	Simazine
25	Chlordane	Butylate

on chemicals that are no longer used in agriculture, many of which are in the deleted list above.

Table 3.3 is a tabulation of insecticide usage according to chemical class in the sixties, "the dawn of environmental awareness," versus the eighties. This tabulation, coupled with the ranking in Table 3.2, gives us an appreciation that we are using different pesticides now than we were two to three decades ago. This was when pesticide debates first occurred and when the majority of today's population was exposed. The use of organochlorines has plummeted while use of their nonpersistent replacements (organophosphates, carbametes, pyrethroids) has increased. So as not to burden the text of this book with a mountain of technical details, I have put a brief description of the toxicology of pesticides in Appendix A for reference. I would urge the reader to consult it for a more complete review of these compounds, both past and present.

It is difficult, based on production and sale figures alone, to get a feel for

TABLE 3.3. Evolution of insecticide usage by chemical class from 1964 to 1982.

Year	Organochlorines	Organophosphates	Carbamates	Pyrethroids	Other
1964	70	20	8	0	2
1966	70	22	7	0	1
1971	45	39	14	0	2
1976	29	49	19	0	3
1982	6	67	18	4	5

Data tabulated as % of total usage.

how much chemical is being used on our farms and whether this has decreased because of these societal pressures. The impact of these processes and mindset can be appreciated by examining Figure 3.1, which shows a continuous reduction in the rate of pesticide application (kilograms per hectare) from 1940 through 1980, a trend that has continued into the present decade. Pesticides that were once applied in pounds per acre quantities now are used effectively in grams per acre. We all experience this when we use the potent herbicide glycophosphate (Roundup®) in our yards. We are using less pounds per unit of area and significantly different ones than we did only a few decades ago.

Silent Spring had a significant effect on decreasing the use of chlorinated hydrocarbons and, whether right or wrong, it singled out this one class of chemicals as essentially evil. Agricultural practices have significantly changed with the introduction of Integrated Pest Management (IPM) techniques that

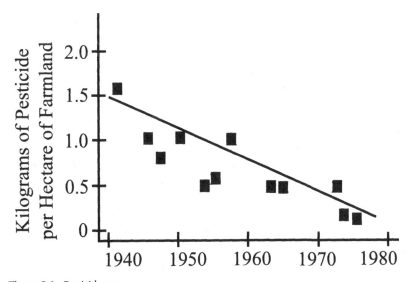

Figure 3.1. Pesticide usage.

are designed to minimize chemical use by combining management and biological and technological strategies to increase agricultural production. There were other reasons that their use and the use of even nontoxic chemicals were reduced. Insects had developed resistance to pesticides much like bacteria have become resistant to antibiotics. Some authors have indicated that this was happening while Rachel Carson was sounding the alarm since compounds like DDT were already in widespread use for two decades. They were easily replaced because other chemicals were already under development to combat resistance. Another reason for lowering use of many modern chemicals is that they simply cost too much! Economics often dictate that costly input—that is, the purchasing and application of chemicals—must be reduced for cost-effective productivity. In contrast, there were some compounds in the past, like DDT, that were so cheap they may have been abused because of this factor alone, possibly resulting in unnecessary ecological impact. We must remember these lessons.

The environmental movement also resulted in the development of organic farming techniques. In these, no synthetic chemicals are used although naturally occurring insecticides, such as pyrethrins, and some acutely toxic ones, such as rotenone, are allowed. Rotenone has recently been reported to potentially be involved in Parkinson's disease in humans based on epidemiological studies and laboratory exposures in rats. I believe the boundary is beginning to overlap between farmers using IPM techniques with minimal insecticides and organic farmers using natural compounds. In large doses, such as encountered by applicators, any pesticides are hazardous to use. It is within the farmer's best interest, both from safety and economical vantage points, to use the lowest amount of the safest compound whenever possible. Finally, the advent of plant hybrids produced through biotechnology have reduced pesticide usage as these crops are now genetically resistant to specific insect pests. This area will be discussed extensively in Chapter 9.

Natural breeding of plant hybrids has been occurring for centuries, and this technique has already produced some insect-resistant produce. However, some of these varieties accomplished this feat by making vegetables, like potatoes, for example, that were more toxic to humans. That is because the process was selected for plants with higher concentrations of naturally occurring neurotoxins and teratogens, such as solanine and chaconine. A similar situation occurred with celery, where abnormally high concentrations of psoralens—a sun-sensitizing group of bitter-tasting natural chemicals—in the modified plants, caused skin toxicity among handlers. Biotechnology, essentially a refinement of hybrid plant breeding, should specifically improve resistance even more and completely avoid direct toxicity to human handlers.

Finally, all agriculture requires fertilizer. Traditional farmers use chemicals such as nitrate. Organic farmers use urban sewage sludge—that is, processed

human feces—or sometimes other sources of animal fertilizer. Although ecologically sound, I wonder what the impact of organic food sales would be if
there were really this level of "truth in advertising." It is time we stop labeling
agriculture purely on the basis of how and what kind of technology was used
to raise the food! Regardless of the way the produce is grown, the final products are equally as wholesome and should be guaranteed as equally safe to
consumers. Where safety is concerned, the properties of the product itself
and not how it was raised should be the primary concern.

In conclusion, our exposure pattern to pesticides is significantly different
than before the appearance of *Silent Spring*. We do not use chlorinated hydrocarbons any longer, and our only exposure to them is as very low residues in
primarily certain animal products, not on produce. We do not spray them on
crops. Only a few percent of the produce consumed by Americans contain
minute amount of pesticides. It is time that people appreciate the real risks
and leave DDT to the historians. Yet from another perspective, we must
remember both sides of the DDT issue when we design realistic regulations
for the next century. Benefits must be a factor, especially when deciding to
remove an existing compound from the market. What are the consequences?

What is the health status of the American public that the alarmists say
have been exposed to pesticides for the last forty years? Are we all dying from
eating our vegetables contaminated with trace levels of these chemicals? This
is the topic of the next chapter.

> Government's convulsions are most dramatic when it deals with
> toxic risks, which former EPA head William Reilly has described as
> regulation "by episodic panic."
>
> (Philip Howard, *The Death of Common Sense*)

Veggie and Fruit Consumption in the Age of the Centenarians

At this point in the book, there may be some readers who persist in being skeptical about the safety of eating produce with even a molecule of a synthetic pesticide present. There is an independent argument that should completely allay this fear. Just how healthy is the American public after consuming pesticides for the last 50 years? Put simply, if we have been the unknowing subjects of a cruel chemical experiment, constantly eating a diet of fruits and vegetables salted with "elixirs of death," then this should be fairly easy to detect. People should be dying in proportion to this intake of poisonous foods. This should be even more pronounced now because our consumption of fruits and vegetables has been steadily increasing. Cancer rates should have increased in the seventies and eighties, a few decades after the period when most of our exposure occurred, since it takes time for cancer to develop. Similarly, all of the other adverse effects of pesticides should have begun to take their toll, showing up as a decreased life span of Americans born and raised in the "age of pesticides."

This is the "acid test" of whether our past practices of pesticide usage actually resulted in widespread harm to our overall health and genetic pool. The prediction made in the sixties was that we would ultimately pay the price.

No longer are exposures to dangerous chemicals occupational alone; they have entered the environment of everyone—even of children as yet unborn. It is hardly surprising, therefore, that we are now aware of an alarming increase in malignant disease.

(Rachel Carson, *Silent Spring*)

This central thesis—that low-level exposure is slowly killing us—is woven through the writings of most pesticide alarmists even up to the present. If we do not die from cancer, then our immune systems will get poisoned, and we will die from infectious disease. The purpose of this chapter is to look at the data some 40 to 50 years later and see if these dire predictions have been fulfilled and to assess the health effects of increased fruit and vegetable consumption.

LET'S LOOK AT THE DATA! WHAT IS THE STATUS OF AMERICAN HEALTH IN 2001?

Epidemiology is the science that looks at populations to determine the presence and incidence of diseases. We examined some ways that human population studies were conducted when we discussed clinical trials in Chapter 1. We should be able to look at data of human death rates and life spans to answer the questions posed above.

There are a few salient points that need to be made. They should be common sense, but often they are ignored. The first is the ultimate truism that has been known and pondered throughout all of history. We are mortal!

> There is a Reaper, whose name is Death,
>   And with his sickle keen,
> He reaps the bearded grain at a breath,
>   And the flowers that grow between.
>
> (Longfellow, "The Reaper and the Flowers")

> For dust thou art, and unto dust shalt thou return.
>
> (Genesis, III, 19)

Logic and reality dictate that we all must die from some cause, be it accident, foul play, or disease. Mortality is the ultimate confounding factor when one tries to determine if exposure to low-level pesticides in our food is causing our demise. Death will always be with us. Our concern is if it comes sooner, or to larger numbers of or in clusters of people, because of exposure to low levels of pesticides in our food. The questions to pose are these:

- Are we dying younger than we were before the age of pesticides?
- Is our life expectancy shortening or is our rate of death increasing?
- Is there any unique disease that can be associated with pesticides that has increased and has been causing our demise?

Now that we are all on common ground, let's look at the data.

The United States Census Bureau tracks our so-called "vital statistics," which include life expectancies and rates and causes of death. Because this data is compiled by census and not for purposes of proving or disproving arguments specifically relating to chemical residues, it should provide us with an unbiased perspective on the health status of the United States. The annual compilation of these statistics will be the primary source of my data. They are readily accessible to anyone who looks (www.census.gov/statab). Let's go to the source. Table 4.1 is an abstract of the life-expectancy data from 1970 projected through the year 2010. In other words, for a baby born in the tabulated year, this is how long he or she is expected to live based on past life expectancies.

Something must be wrong! This data clearly shows that our life expectancy is actually *increasing*! Similar patterns have been noted with *worldwide* life expectancies at birth, which also increased from 48 years in 1950 to 64 years in 1990. Whatever statistics one chooses, the same conclusion is always made. Our life expectancy has been steadily increasing. This has been having a significant impact on the cost of health care as there is an increasing percentage of the older population due to increased longevity. There are more centenarians (absolute numbers and percentage of population) alive today than ever before in history! Some demographers feel that the maximum average life expectancy for humans as a species may be 85 years. Thus the data in Table 4.1 indicate that we are approaching this mean barrier.

Well, let's look at the flip side of the coin and assess death statistics. Let's examine the relevant data, which is death rate expressed per 1,000 population, and the same figures adjusted for different age groups, to determine if the death rate is increasing with time (see Table 4.2). These data are best expressed as fre-

TABLE 4.1. Abstract of life-expectancy data from 1970.

Decade	Life Expectancy at Birth
1970	70.8
1980	73.7
1990	75.4
2000	77.1
2010	78.5

TABLE 4.2. Death rate figures adjusted for different age groups.

Year	Death Rate	<1	1-4	5-14	15-24
1980	977	1,429	73	37	172
1990	918	1,083	52	29	147
1998	876	822	38	23	118

Death rates per 100,000 population.

quencies per affected population because percentages can be very misleading. In addition to looking at the overall rate, we will also look at some younger people who often are singled out as having higher exposure to pesticides.

Here it goes again: in all categories, the death rate is decreasing! These years are particularly relevant because they bracket the time that our population was exposed to pesticides. All figures indicate that our life expectancy is increasing and death rate decreasing over time. If we extended these tables back to any other earlier time, the identical pattern is seen. Similarly, we can also examine infant mortality rates and causes of infant death over long time periods and all are decreasing. Worldwide, the infant mortality dropped from 155 deaths per 1,000 live births down to 70 deaths per 1,000 live births from 1950 through 1990. This further supports the fact that there has been an overall improvement in human health. There is no data to support the contention that we are being poisoned.

The final vital statistic to examine before we can rationally discuss this issue relates to what is the cause of death. As with the preceding data, this is expressed as deaths per 1,000 people. Inspecting the raw data from which these compilations have been reduced is an education in itself because one can appreciate the numerous *preventable* factors, such as motor vehicle accidents, homicides, suicides, etc., that are killing us. The general rate of infectious disease—the primary cause of death until the advent of sanitation early last century and antibiotics in the 1950s—has decreased over the decades with the exception of the 1990s when we begin to feel the influence of AIDS. This dreaded disease, and what I believe this trend is really showing, will be fully discussed at the end of the book in Chapter 11. However, for our present purposes, I have only tabulated the major causes of disease that could be contributable to chemical causes (see Table 4.3).

Now the data starts to get interesting and requires some interpretation. First, recall that these people who are dying are doing so in smaller numbers and at older ages. This can be seen from the previously tabulated death rate and life expectancy data. The primary killer remains heart disease. This is not unexpected as we are talking about the cause of death for people in the last

TABLE 4.3. Major causes of disease that could be contributable to chemical causes.

Year	Heart Disease	Cancer	Lung Disease	Liver Disease
1970	5.0	1.6	0.15	0.16
1980	4.4	1.8	0.25	0.14
1990	3.7	2.0	0.35	0.10
1998	3.5	2.0	0.42	0.09

Death Rates per 1,000 people.

decades of their lives. The next killer is cancer, which does show a mild increase that will be fully discussed below. It is interesting to note that the rate from liver disease is actually decreasing, the complete opposite of the prediction in *Silent Spring* that:

> In connection with the nearly universal use of insecticides that are liver poisons, it is interesting to note the sharp rise in hepatitis that began through the 1950s and is continuing a fluctuating climb.

The major cause of chronic liver disease today, whose rate is *decreasing* over the last 20 years, is cirrhosis secondary to chronic alcoholism. The rates of viral hepatitis are relatively stable, and there is no evidence of a pesticide-induced epidemic of hepatitis.

This example is a good one because it shows how misleading the wrong analysis of these same sets of data can be if one looks at the tabulations in the wrong way. One cannot use absolute numbers of deaths to assess our health since they will always be higher if the total population is higher. Unfortunately, this frequently occurs in many fields. For example, more people die in New York City than do in my hometown of Raleigh, North Carolina, simply because New York's population is over 8 million while Raleigh's is only 1 million. Similarly, in 1950, the population of the United States was about 150 million. In 2000, it was almost 275 million. More people died in 2000 than in 1950. If two people in every thousand (average rate) died from cancer in both decades, then 300,000 would have died in 1950 and 550,000 in 1990. More people died of cancer (and any other disease category) primarily because there were more people to start with. The slight increase in rate—from 0.16 to 0.20%—is greatly magnified when absolute numbers are used. I am astounded to read in some popular texts a conclusion similar to the preceding, which serves as "evidence" of a "cancer epidemic" since total numbers of cancer deaths are increasing. *This is wrong and should not be tolerated.*

The cancer rates actually are more complicated than this because there are deaths from all types of cancer. Another factor is that if one examines age-adjusted cancer rates, as expected, there is a continuous increase with age no matter what category one chooses to examine. For example, the 1998 rate of cancer deaths per 100,000 is tabulated in Table 4.4.

TABLE 4.4. **1998 rate of cancer deaths per 100,000 people**

Age Group	Males	Females
15-24	5.2	3.7
25-44	23.6	28.3
45-64 *	260.1	216.5
> 65	1,424.9	913.5

* Cancer is the leading cause of death in this age bracket.

Note the dramatic increase for people older than 65. We are dealing with a disease that affects the aged. It is interesting that if this analysis were done 100 years ago, this category would almost be nonexistent because few people lived that long and cancer was never diagnosed as it wasn't discovered! The primary cancer with which we are dealing is lung or respiratory cancer through about 70 years of age, after which cancer of the digestive and reproductive systems start taking their toll. In a 1988 study by the National Cancer Institute, using data in addition to these simple census figures, the age adjusted mortality rate for all cancers (except lung) has been steadily declining since 1950, except for those individuals greater than 85 years of age. One has to die of something, and cancer will play the role of Longfellow's Reaper for many decades to come. As mentioned earlier, 85 years is a benchmark age after which decreases in mortality seem difficult to lower for the population as a whole. However, some individuals do beat the odds as evidenced by the population explosion in centenarians.

There is uniform agreement that the primary cause of cancer in the United States is lung cancer due to smoking. If one breaks the preceding cancer rates into types of malignancies, there has been a steady increase in lung cancer for all categories tabulated. Smoking (and thus lung cancer) is the reason for the slight increase noted in total cancer rates. Consistent with this, the rate of lung disease, categorized as "chronic obstructive pulmonary diseases and allied disorders," is increasing. If lung cancer is removed from the tabulation, the rates of cancer stabilize or decrease as in the NCI study. I could go on and on comparing rates of different types of cancers—for example, increased colon cancer but decreased stomach cancer. In 1993, the 4 major causes of new cancer, accounting for 55% of the total cancer, were lung, breast, colon, and prostate. I do not feel that this continued type of analysis would add anything new to our discussion. There is no cancer epidemic!

There have been numerous epidemiologic studies specifically designed to probe the incidence of cancer by taking into account patterns of *smoking* and *chronic alcohol* consumption. Study after study indicate that these two factors significantly increase the rate of cancer in our society. There is also consensus that our rates of skin cancer and melanoma are increasing because of exposure to ultraviolet radiation primarily from the sun. The public widely acknowledges that asbestos is a major cause of cancer, and this is true. Occupational exposure, such as manufacturing, mining, or installation of a specific filament type of fiber, is a risk. Ship builders in World War II were at risk. However, the data shows that the significant factor which must be present for the most common type of asbestos-induced lung cancer to occur— interstitial fibrosis—is concurrent smoking. The other type of cancer— mesotheliomas—will occur typically with occupational asbestos exposure. These types of interrelationships hold for numerous other cancers and toxicologic problems. Recall that the liver is a primary defense organ that metabolizes and eliminates foreign chemicals from our body. If one destroys the liver by chronic alcohol consumption, it cannot perform its function and consequently is more susceptible to other insults if one manages to survive the primary disease.

Some may argue that the reason our life expectancy is increased is not because of a decreased incidence of disease but rather because the miracles of modern medicine can now cure disease and postpone death. This would suggest that, contrary to all the data above, the incidence of cancer is actually increasing, but modern medicine has "found the cure," and thus we are no longer dying from it. There are two components of this hypothesis which must be addressed. First, modern medicine has made significant progress in eliminating most causes of infectious disease (except for AIDS) and in taming heart or cardiovascular disease and its complications, for example hypertension. This has allowed people to live longer and get more cancer, as fundamentally, cancer is often a disease of aging. The body's ability to regulate cell growth is lost, and cells grow unabated. Unfortunately, the second part of the prowess of modern medicine is not true. We have not cured cancer. We have made some excellent progress in curing a number of rare childhood malignancies such as leukemia. We have saved lives by earlier detection. This also implies that we are better at diagnosing cancer than in years past and thus should be finding more, which by itself would increase cancer rates.

If we return to the Census Bureau for data, we can look at a comparison of 5-year survival rates by cancer type from the seventies through the nineties. Unfortunately, these have remained flat for this time period. There has been marginal improvement in some types; however, the "Reaper" changes his scythe and ultimately attacks anew. Thus there is *no* evidence that cancer death rates have stabilized solely because of an increase compensated by improved survival and therapy.

Dennis Avery estimates that less than 3% of all cancer deaths are caused by *all* the combined forms of environmental contamination and pollution. Both Avery and Bast and coauthors cite work by Dr. Robert Scheuplein of the FDA who specifically looked at the incidence of pesticide-induced cancer in relation to the total amount of cancer attributed to food. At the most, pesticides could account for 0.01% of food-related cancers, a topic we will address extensively in the next chapter. Many authors cite similar conclusions based on an often-quoted landmark epidemiological study on the causes of cancer, done by Doll and Peto in 1981 and published as the book *The Causes of Cancer*. The authors studied cancer deaths in our country from the 1930s to the late 1970s, the precise period when the risk of pesticide-induced cancer would have been expected to peak as chemicals such as DDT were in heavy use during this period prior to the influence of *Silent Spring*. They concluded that 98–99% of cancer in the United States is due to either smoking, genetics (something we as of yet have no control over), or diets high in fat and low in fiber. That leaves a vanishingly small amount of cancer that could remotely be due to any type of chemical contamination. There is not a pesticide-induced cancer epidemic.

Avery uses another poignant example of the lack of cancer risk by examining data from a segment of the population that should have higher rates if pesticides truly cause cancer. These are the farmers and manufacturers who are *occupationally* exposed to higher pesticide levels. Part of these studies have been prompted by recent public concerns about the safety of the number one pesticide used in this country, atrazine, the compound that has occupied the top of the chart (Table 3.3) for the last few decades. Some high-dose animal studies suggest that atrazine may be associated with mammary tumors in one strain of susceptible rats. We will deal with this issue for other pesticides in far more detail in Chapter 7. However, since we are looking at cancer statistics in general, this is an opportune time to look at one more.

Atrazine has been extensively used for over 30 years and is even present in trace levels in our water supplies. If this compound is toxic, we should be seeing some effects. Based on our earlier analyses, we are not seeing it expressed in our national vital statistics. Thus let's look at the incidence of breast cancer in women who are exposed by being involved in farming and forestry. These women have a *lower* incidence of breast cancer—that is, only 84% of the average, for American women. Similarly, workers manufacturing atrazine have disease incidences the same as the general population. When the EPA reviewed its own laboratory animal studies for cancer, it *raised* the "no-effect" level ten-fold, indicating that it was ten times safer than previously thought. When all of the numbers are "crunched" through the risk assessment formulae, taking into account parent compound and its breakdown products, this would raise the violative concentration in drinking water from the current 3 parts per billion to 20 parts per billion! For an average woman living in a city to achieve this no-effect level, she would have to drink 154,000 gallons of water per day! Again, that would only put her at the *no-risk* level. Of course, since this level of water consumption greatly exceeds water's own "no-effect" level, our subject would have died from water toxicity—drowning—long ago.

There is no clear evidence that adverse effects to pesticide consumption are occurring, even in populations with higher exposure. This is another example that shows there is essentially no risk, and we should move on to other ways to spend our money. However, there remains a movement afoot to ban atrazine.

This lack of evidence of pesticide effects, based on historic data when pesticide use was greater than today (and even in exposed populations), does not warrant the current public concern. In fact, we will now examine evidence that we should be eating more produce to prevent us from getting cancer by all the other nonpesticide causes to which we are exposed.

VEGETABLES AND FRUITS ARE GOOD FOR YOU!

I hope you are convinced that our health is improving, and while this data is fresh in your mind, let me argue that part of the reason for this "epidemic of good health" is eating fruits and vegetables. Back to the Census Bureau for real and current data. From 1970 to 2000, the United States annual per capita fruit and vegetable consumption has increased by some 25% to 30%. This parallels the total produce consumption data documented in the previous chapter. This is the same time period during which our life expectancy has made its substantial increases and is the very same produce which was surveyed and analyzed for trace pesticide residues. *The data strongly suggest that increased fruit and vegetable consumption, even with those few molecules of nasty pesticides present, improves our health.*

Making such correlations can be fraught with trouble because in most epidemiological surveys a relationship alone does not prove causation. However, I would now like to spend the rest of this chapter precisely illustrating, using studies designed for this purpose, that there is significant data to support the concept that the more produce one eats, the *less* chance of cancer and the *greater* your life expectancy.

You have now entered a "logic warp." If you search the literature using key words for pesticides and health, you find that the question people try to answer is whether the presence of pesticides causes cancer and other adverse health effects. If you search for pesticides and vegetables, you get the kind of data tabulated in Chapter 3—incidences of trace residues on produce. If you search for vegetables and health, you will get the rest of this chapter—that is, eating veggies is good for you. The "logic disconnect" is that the primary source of exposure to pesticide residues in our diet is on fruits and vegetables. Assessing what happens when you eat produce, which was shown in Chapter 3 to have trace (nonviolent) residues present in one-third of samples, is identical to assessing the health effects of consuming pesticide residues. As I will demonstrate, eating produce is healthy. Why isn't this connection more widely acknowledged?

I was first impressed with the *strength* of this relationship using a technique that I often employ when approaching any new area of science. I go to the library and seek out a recent review article on the subject and then chase down the citations in the bibliography, so as to read the evidence for myself. What I am looking for is a wide consensus, represented by articles published in a large number of *different* types of journals. As a practicing scientist, I am fully aware that strange things can often happen in the laboratory; however, it is rare that they can be reproduced. *Reproducibility and consensus, although not perfect, are the cornerstones upon which modern science and medicine are built.* There is clear evidence that consumption of fruits and vegetables reduces the incidence of cancer and may even be protective against lung cancer due to smoking. Individuals with diets *low* in fruits and vegetables consistently show *higher* incidences of lung cancer. Dr. Bruce Ames indicated that of 172 studies he reviewed in the literature, 129 of them showed a significant *protective* effect of vegetable consumption on lowering the risk of cancer. The remaining studies were neutral on this point. The produce consumed in all of these studies were off of the grocer's shelf and were *not* organically grown or selected by any criteria related to pesticide levels. This evidence is accepted by scientists with widely different areas of expertise—nutritionists, epidemiologists, physicians, chemists, biochemists, toxicologists, etc. These are not isolated conclusions but as close to scientific fact as we will ever get. They are consistent with our previous analysis of vital statistics, which show increased life spans in a population which has been increasing its consumption of fruits and vegetables over the last few decades.

The popular press also reports on the benefits of fruit and vegetable consumption in promoting health. Flavenoids (the yellow pigments) found in citrus fruits, indole-3-carbinol and sulforaphane in broccoli, as well as several phytochemicals in tomatoes, have been reported to help fight cancer. As fully discussed later, the benefits of eating the whole fruit or vegetable, versus just a supplemental pill, were advocated. The original literature was properly interpreted. What fascinates me is that in articles that push the benefits of fruits and vegetables, pesticide residues that "come along for the ride" are never mentioned. In contrast, when reports are written about pesticide residues or environmental chemicals, no mention is made of the benefit of produce consumption studies showing the dramatic health benefits. Although the focus of some of these may be on environmental or ecological issues, it is the potential human health risk that is often touted as the emergency. Obviously the connection is not being made.

Even in an organic guidebook to pesticides published for the Rachel Carson Council, the following statement is made:

> Carcinogens occurring naturally in plants (e.g., aflatoxins) may not be ignored, but it is well known that vegetarians have a considerably lower cancer risk than do omnivorous humans.

Vegetarians do not select only produce grown pesticide free. Like all other produce consumers in these studies, most eat "off the shelf" produce. Therefore, logic dictates vegetarians will intake more pesticide residues. Yet they live longer! Another major reason that vegetarians may be living longer is related to their reduction in animal fat intake, which will reduce their incidence of heart and other diseases. However, their intake of pesticide residues on produce should also be higher, and thus these studies refute the link of pesticides with increased disease.

The focus of current research is to identify the natural factor in fruits and vegetables that protect against lung cancer so that supplements can be given. The hope is that if smoking cannot be eliminated, maybe dietary supplementation with specific nutrients will offer protection. This is where the scientific controversy lies. If one measures beta-carotene concentrations in blood, there is often a correlation to the anticancer effect. The carotenoids are a group of natural plant hydrocarbon nutrients and include beta-carotene, alpha-carotene, beta-cyrptoxanthin, lutein, and lycopene. They are converted in the body to retinol (vitamin A). They are found in dark-green and yellow-orange vegetables. However, some studies of artificial beta-carotene supplementation—that is, beta-carotene obtained by means other than eating vegetables—fail to show the protective effect. Therefore, the beta-carotene level correlation may just be secondary to high levels of fruit and vegetable consumption, which correlates to the protection. Other phytochemicals in produce may actually be responsible for the protection. Beta-carotene may only serve as a marker of vegetable consumption. Again, correlation and causation are two different entities.

There are hundreds to thousands of such potent chemicals in produce. In addition to vitamin A, interest has focused on vitamin B, especially folic acid, in preventing birth defects and recently even heart attacks and strokes. There is also interest in the following:

- Vitamin C (the fruit connection) protection against mouth, throat, and stomach cancer
- Vitamin D's effects on preventing breast, bowel, and prostate cancer
- Vitamin E's reducing the risk of heart disease
- Several minerals, including selenium, reportedly protecting against cancer

Even the fiber found in plants helps protect against colon cancer. With the advent of sophisticated analytical techniques, many other additional chemicals are being identified (see below) which will probably play a more important role in future research. Breakthroughs in our understanding of genomics are developing plausible mechanisms by which these beneficial effects may be occurring. Similar "hard science" could be quoted to support numerous other beneficial studies showing how produce protects against cancer. We discussed the carotenoid connection with lung cancer because it is the tightest data and illustrates the most benefit. A similar argument could be made for the benefits of folic acid. The link between vegetable consumption and health is just as strong, but we just don't know what it is in the vegetables that protects us.

Broccoli has been touted to protect against the formation of breast tumors, possibly because of its indole-3-carbinol derivatives. Broccoli also contains sulforaphane, which aids cells in excreting carcinogens and other chemicals that prevent binding of carcinogens to DNA. Indole-3-carbinol is an antiestrogen that decreases the likelihood of laboratory animals developing breast tumors. Consumption of cabbage has been shown to decrease the risk of colon cancer. Phenethyl isothiocyanate, which despite its name is a natural chemical, is a constituent of cabbage and turnips which has been shown to have some protective effects against lung cancer by blocking metabolic activation of a carcinogen in tobacco smoke. Even nitrates have been implicated as being protective against lung cancer. The list is almost endless as to the chemicals being identified on almost every fruit or vegetable being studied. Some of these will be mentioned in Chapter 5 when we discuss the organic chemical consumption of the "natural" foods we eat.

The purpose of this book is not to identify the relationship of vitamins to cancer; rather it is to show that eating vegetables does not cause cancer. It seems ridiculous that this point needs to be made. All of the studies indicate a protective effect and thus offer no support to the contention that eating produce with trace pesticides is harmful.

Let me illustrate how dangerous simple correlations can be. If I wanted to play the "pseudoscientist game," I could make the observation that since increased produce consumption is related to decreased risk to cancer (as was shown in the beginning of this chapter), then maybe it is the trace amount of pesticides that we know are on produce (which was shown in Chapter 3) that provides the protection against cancer! This is consistent with the lack of effect with artificial supplementation. If you look at these data, the evidence is significantly stronger than those who argue that eating produce with these trace pesticide levels should be detrimental. They only use "threat" of a potential disease to strike fear in people's minds. Ignore the fact that there is no data to support this. What is lacking in my protective argument is a mechanism of trace pesticide effects and the fact that we are not dealing with a single pesticide. All of the arguments which I presented against the adverse effects, I must also present against my own "straw-man" hypothesis because it would be essentially impossible to prove, even with the data to support it.

Since I began paying attention to these issues in the mid-1990s, there has been a group of investigators, led by Dr. Edward Calabrese of the University of Massachusetts, who have developed a hypothesis that low-dose exposure to chemicals, including pesticides, at levels below toxic thresholds may actually have beneficial effects. This hypothesis, termed *hormesis*, postulates that properly designed studies show a beneficial biological effect at concentrations below toxic or "no-adverse-effect" thresholds; that is, there exists a "U-shaped" dose-response relationship. This low-dose stimulatory effect is usually no more than two-fold greater than control responses, but may be due to chemical-induced effects which improve homeostasis. The body thus "over-compensates" for low chemical exposure. Hormesis could be a plausible explanation for the lack of adverse effects seen with low-level chemical residue consumption on produce. At the very least, it challenges low-dose extrapolations of adverse effects including carconogenesis from high-dose toxicology studies.

In fact, if certain anti-pesticide groups were to have their way, we could lose a lot of our progress in health care, because chemicals which allow this level of food production to occur could be removed from the market. Fungicides such as captan, which kill aflatoxin-producing mold, would be banned. Then we would be stuck eating food contaminated by a much more potent chemical which is *known* to cause cancer in animals and man. Crops which are wet also tend to have increased mold problems, and thus with recent flooding of our farm areas, we could be putting ourselves at significant risk even while we have the technology to eliminate it.

The conclusion of all of these data is that vegetables and fruits are good for us. The epidemiological data support this contention. This is the same produce that has been sampled by the FDA and found to contain trace residue levels of almost 100 pesticides. The fact that these same vegetables contain trace levels of pesticides indicates that a diet of fruits and vegetables on the whole is extremely beneficial to one's health. The definitive human trials—in fact some 170 of them—to assess the health effects of trace pesticide consumption have been conducted. The results are the same. *Fruits and vegetables are good for you.* As Mom—who was 88 at the time I wrote this book—has always said:

Eat your vegetables!
An apple a day keeps the doctor away!

5

Chemicals Present in Natural Food

The reader should now feel comfortable that eating fruits and vegetables is good for you and that even if they contain the trace levels of pesticides documented in Chapter 3, the risk must be minimal. Now I will present you with the ultimate paradox. Up until this point in the book, we have completely focused on the theoretical hazards associated with artificial man-made chemicals in our food. Implied in many discussions of food safety, and especially chemical residues, is the concept that if the food is natural and not tainted with man-made chemicals, then it must be safe and healthy. This is deeply rooted in the nostalgic view of the benevolence of nature.

> Nature never did betray
> The heart that loved her
>
> (William Wordsworth, "Tintern Abbey")

If Mother Nature made it, then many believe that harm cannot come their way. This concept is supported by the astronomical growth of the natural foods and supplements industry. Interest in herbal and oriental medicine continues to surge. Although this romantic belief is generally true, modern man has forgotten some of the finer points inherent to this rubric that our ancestors regularly acknowledged. They did this because they knew nature is a constant foe.

> Are God and Nature then at strife?
> That Nature lends such evil dreams?

So careful of the type she seems,
So careless of the single life.
(Alfred, Lord Tennyson, "In Memoriam")

Plants have been used as poisons since ancient times. The poison of choice in ancient Greece, and the one used to kill Socrates, was a potion made from poison hemlock. In fact, poison hemlock (*Conium maculatum*) contains the toxic chemical *coniine*. It was one of the first alkaloids discovered in 1827 and the first to be prepared synthetically by Ladenburg in 1886. It was the use of crude plant preparations, which had unknown potencies and contained natural contaminants, that heralded the modern pharmaceutical industry. Medicine and toxicology have always been intertwined, and its practitioners have rarely judged the "value" of a chemical solely on the basis of whether it was of natural or of synthetic origin. The fact is that most of our synthetic drugs have their origins in natural compounds. It is only recently in the popular press that compounds, even if they were originally isolated from plants, have a stigma attached to them if they are presented as products of modern chemistry. I would even go beyond this and state that some in society do not necessarily have a phobia for chemicals but rather for chemical nomenclature. This will become increasingly evident as we examine what the names of the chemicals are that actually compose the food we eat.

What never ceases to amaze me is that much of the current hysteria associated with promoting natural products and avoiding synthetic chemicals would put medicine back to the state it was in the eighteenth and nineteenth centuries, before synthetic drugs or pesticides were invented. As I will discuss, plants generally contain a multitude of chemicals, some beneficial and some not. Their concentrations are dependent upon a host of agricultural factors—temperature, water, humidity, nutrient source, etc.—that make it difficult to predict in advance what their "strength" will be. Most of the chemicals that we so boldly label as good (vitamins) or bad (poisons, carcinogens), relative to how they affect humans, are integral to the plant's survival. Some may be synthesized only when the plant is stressed and must be more competitive in a specific environmental situation (e.g., drought) for its own survival. As will be shown, many of these chemicals are known carcinogens that are then eaten along with the rest of the plant. Modern pharmacology advanced beyond this primitive state decades ago by identifying and selecting the "good" chemicals and removing them from the milieu that contained the natural poisons. This allowed for definition of standards of purity, efficacy, and toxicity.

The fundamental problem facing us in this book is that the political and legal regulatory systems that developed are completely focused on synthetic chemicals. These products of modern technology must be scrutinized by the most advanced science since the dawn of civilization so that the public health is not damaged. If the product is natural, there is no regulation, and our citizens can consume tons of unlabelled natural carcinogens.

This was driven home when Congress legislated a few years ago that close regulation of natural nutritional substitutes, so-called nutraceuticals, were off-limits to FDA unless very specific indications of adverse effects were reported to the agency. Thus manufacturers were clear to market large quantities of "natural" herbal remedies to the public without clear-cut evidence of either safety or efficacy. The most obvious example of a potential toxicant being marketed are those diet pills containing ephedra alkaloids, obtained for example from "ma huang." In certain individuals, tachycardia and ventricular arrhythmia may occur after ingestion of large quantities. This has resulted in removing over-the-counter preparations from the market (diet pills, some decongestants), yet "natural" herbal products containing chemicals with the same mechanism of action remain available.

The chemistry of the compounds produced in these natural laboratories (plants) easily matches the sophistication of any modern synthetic laboratory. Plant cells are capable of producing a wide variety of organic compounds. They may be so sophisticated that modern science, in the guise of biotechnology, is essentially using the plant's own chemical manufacturing machinery when it dictates the "working orders" through control of the "specifications or recipe," DNA. This is the essence of genetic engineering. We are using plants to make our chemicals. In North Carolina, where tobacco once was king, scientists now have manipulated the genetics so that the plant may synthesize insulin. Instead of producing drugs in large chemical factories, farmers someday will be harvesting complex chemicals by the acre! The line between synthetic and natural is getting blurrier and blurrier each day.

One of my favorite examples I use to illustrate the stigma attached to chemical exposure involves some members of chemical classes that contain the cyanide grouping (CN). The reason I pick this compound is that most people rightly associate cyanide as being a potent poison. Generally, these are inorganic cyanide gasses such as hydrogen cyanide (HCN) or its water soluble salts such as potassium cyanide (KCN). The potency of this chemical was tragically demonstrated when methyl isocyanate was released in the explosion of the chemical plant in Bhopal, India, in 1984. Thus cyanide is rightfully associated in most minds with poison and danger.

Considering this background, would you eat a product if the label listed the following chemical ingredients: *allyl cyanide, 1-cyano-3-methylsulfinyl-propane, caffeic acid, allyl isothiocyanate, and 4-methylthiobutyl isothiocyanate?* In total, this product contains almost 50 additional sinister-sounding chemicals. These ingredients are present in concentrations up to 100 to 1000 times greater than those of the pesticide residues discussed in Chapter 3. The National Institute of Environmental Health Sciences' (NIEHS) National

Toxicology Program has determined that some of these chemicals are carcinogenic in rats.

Now you are probably thinking that since this product contains such nasty ingredients, documented by scientific studies to cause cancer in laboratory animals, it must no longer be on the market. Maybe it's a rat poison. Surprise! This product is organically grown, 100% natural cabbage taken right from the local health food store. This is the same cabbage that was discussed in Chapter 4 as possibly being protective against colon and lung cancer. I believe that the uneasiness that most readers would feel when reading this hypothetical "label" is a clear illustration of our phobia for chemical nomenclature. If this were a synthetic chemical, the media may have reported your consumption of this "product" as being part of a long-term human experiment, testing the effects of lifetime consumption of the rat carcinogen *allyl isothiocyanate.* However, since the "product" is coleslaw and the "sponsor" is Mother Nature, there is no story.

A similar argument can be made for numerous fruits whose seeds contain acutely toxic doses of cyanide compounds. One example is the apple, even without Alar® or other pesticide residues included. Like cabbage, apples also contain the known carcinogen *caffeic acid.* Pitted fruits, such as apricots, peaches, and sweet almonds contain the cyanogenic glycoside, *amygdalin.* When was the last time you picked up an apple or apricot that contained a warning label indicating that the fruit contained cyanide, a chemical known to be a *human* poison? Amygdalin is the primary component of the fad anticancer drug, Laetrile, which was administered by injection but was poisonous to children who ingested it.

Let us assume that you are a new biotechnology company and decide to synthesize the apple from "scratch" and get regulatory approval for its distribution. You would first have to identify all chemical components and then test each individual component for toxicity, using the tools of modern toxicology to prove safety to the appropriate regulatory agency. Assuming that you had enough money for this daunting task, our old apple would never be approved because some of the chemicals would be carcinogenic when tested in rodents. Also, there would be a risk of acute poisoning since many rats would have died early in the high-dose studies from cyanide toxicity. If it could be approved, which I doubt, imagine the label and safety warnings attached to it. Figure 5.1 illustrates a possible label. We should all be thankful that the apple is a product of evolution and thus does not require regulatory approval.

This discussion illustrates the *central dilemma* facing scientists, regulators, and the general public. Many of our foods contain chemicals that should be harmful to us, yet we have evidence that the foods themselves are actually beneficial! The chemicals occur at concentrations that are within the acutely toxic range, yet the foods are still allowed in the marketplace without labeling. Dr.

Figure 5.1. Apple label.

Bruce Ames is a California biochemist who invented the Ames test for chemical mutagenicity. Recently, he has been a prolific author on the relative risks of consuming natural versus synthetic chemicals. He has estimated that 99.9% of all pesticides in our diet are natural plant pesticides, which occur in concentrations ranging in the parts per million category rather than in the parts per billion category, as seen with synthetic pesticides. In contrast, trace levels of synthetic additives, which only may be suspected of being toxic in studies where animals are "megadosed," are banned from the grocer's shelves. In this latter case, it is conceivable that a manufacturer will be sued if these pesticides were found at violative levels in our food. Will I be taken to court because I feed my children cabbage or apples containing multiple natural carcinogens? Labeling food as "natural" and "pesticide and additive free" implies that the food is free from harmful chemicals and safer than produce grown without synthetic pesticides. However, the most potent chemicals are the natural components of both cabbage and apples. Is this policy correct or even ethical?

This discussion will be revisited at length in Chapter 8, when we discuss the case of BST (bovine growth hormone) in milk. This is essentially nontoxic to humans but has been stigmatized in the popular press because it is a product of modern biotechnology. There are other natural hormones in food with which we should be more concerned. Again, in keeping with my labeling concerns from above, court cases are being argued that milk should not be labeled "BST Free" since it implies that it is safer than milk from cows given BST when there is no objective scientific evidence that they are different.

There are numerous examples of other natural toxins in food. This whole book could be a tabulation of hazardous materials found in everyday food. Parsley and celery contain up to 30 parts per million *methoxypsoralen*, another rodent carcinogen. Orange and mango juice have up to 40 parts per million of the carcinogen *limonene*. Our old friend, the carcinogen *caffeic acid*, is a natural constituent of a dozen different spices, as well as being found in produce ranging from apples, grapes, and pears to lettuce, carrots, and potatoes. A separate volume could be written on the dangers of spices alone, most of which contain known carcinogens. For example, basil and fennel each contain over 3,000 parts per million of the rodent carcinogen, *estragole*, while nutmeg contains as much *safrole*. Black pepper contains both *safrole* and *limonene*. I will finish with a hot cup of roasted coffee, which in addition to containing *caffeine*, also contains *caffeic acid* and *catechol*. Throughout history our civilization has been, and will continue to be, constantly exposed to a multitude of natural carcinogens in all of our produce and spices, at concentrations thousands to millions of times higher than any artificial chemical!

The point of this book is not to catalogue every natural carcinogen known to modern science. If that is the goal, the interested reader should consult the exhaustive literature by Dr. Ames listed in the references. My point is that these known toxins are widespread in our produce yet we already know that eating these same fruits and vegetables is actually good for your health. Why?

One excellent food that illustrates this dichotomy is the peanut. First, there is a well-documented physical hazard associated with children who choke on them. Should we label peanuts to protect us from these acute hazards, which probably killed more children than exposure to any trace level of pesticide on the same peanut? Secondly, some individuals are acutely allergic to peanuts or any food containing them. This is a serious problem for those sensitive individuals, and efforts continue to be made to label foods containing peanuts to alert these individuals. The main toxicological problem of peanuts and products made from them, such as my personal dietary staple, peanut butter, is the problem of contamination with molds. *Aflatoxin*, found in moldy peanuts, is the most potent human carcinogen known and is well documented to produce liver cancer.

To eliminate moldy peanuts, synthetic fungicides are used. The scientific, ethical, moral, and legal dilemma facing us is that many effective fungicides test positive for inducing cancer, whether the tests are done in laboratory animals or *in vitro*. Some of the most widely acclaimed tests use bacteria to determine if the chemical can damage the bacterial DNA. An effective fungicide is usually toxic to fungal cells and the relatively closely related bacteria, so it is no surprise that these potent chemicals test positive in these tests when the chemicals are given at high doses. Since they are synthetic chemicals, they are banned from use. Banning fungicides does remove this chemical carcinogen, but it also allows mold to grow and produce mycotoxins such as the carcinogenic aflatoxins. "Mother Nature" does not have to get FDA approval to evolve plants that produce carcinogens such as aflatoxin. Are we safer or more at risk because of this policy?

I should point out that many of the earliest fungicides were indeed hazardous to human health and were only developed because there was a positive benefit to risk ratio—that is, they did more good than harm. These included chemicals whose main toxic ingredients were mercury or arsenic. As discussed in Chapter 3, early pesticides containing these toxic metals were probably the culprits partially responsible for some of the ecological problems discussed in *Silent Spring*. It is worthwhile at this point to briefly review the toxicology of the mycotoxins, since many were first described because they caused disease in humans and were the impetus for developing chemical fungicides. What was the trade-off?

Mycotoxins are biologically active chemicals produced by various species of fungi and molds. The molds that produce mycotoxins are often found as common contaminants of grains or grasses. The production of mycotoxins by these organisms may only occur under specific environmental conditions, such as defined temperature or humidity, or after mechanical damage. When farming or storage conditions are not optimal, mold may form. This was a major historical problem that was largely eliminated by the development of fungicides. Certain species of fungi most often produce mycotoxins in stored grains, whereas other fungal species most often produce mycotoxins in plants that are in various stages of growth. As a group, mycotoxins represent some of the most potent toxins presently known to modern toxicology. Many are potent carcinogens, mutagens, and/or teratogens, with aflatoxin being one of the more potent carcinogens known. These agents have even been suggested as agents of chemical warfare or bioterrorism (see Chapter 10). These are not sinister chemicals synthesized by man but are natural components of our environment.

For regulatory purposes, mycotoxins are considered contaminants, whereas fungicides and pesticides are considered indirect food additives. Under the Delaney Clause of the Federal Food Drug and Cosmetic Act, "no residue" of any food additive that has been demonstrated to be a carcinogen is permitted in foods intended for human consumption. The Delaney Clause does not apply to contaminants, so low levels of mycotoxins, therefore, are permitted in food even when the mycotoxin (e.g., aflatoxin B_1) is known to be a potent carcinogen. Chemical pesticides and fungicides, although of significantly less toxicological potential, are automatically banned because of their synthetic origin. Their benefit of decreasing exposure to the known toxic contaminants is not factored into the equation.

The aflatoxins are exceptionally potent metabolites of the fungi, *Aspergillus flavus* and *Aspergillus parasiticus*. In the United States, aflatoxin contamination most commonly occurs in corn, cottonseed, and peanuts. Contamination of grain by aflatoxins affects many regions of the world, including the United States. Aflatoxin contamination is most prevalent in the South, where the warm temperatures and high humidity of this area provide a favorable environment for mold growth. Infection of grain by aflatoxin-producing fungi can occur any time during the growth, harvesting, transport, or storage stages.

There are a number of other agriculturally significant mycotoxins that produce a public health risk. *Fusarium moniliforme* contamination of corn produces the mycotoxin fumonisin, which may cause disease in both animals and man. This mycotoxin may also contribute to human esophageal cancer in certain regions of the world. The ochratoxins represent a family of mycotoxins produced most often by *Aspergillus ochraceus* and *Pencillium viridicatum*. Ochratoxins occur as natural contaminants in a number of feeds including corn, peanuts, barley, oats, beans, and hay. Ochratoxin-producing fungi generally thrive in warm, moist areas. Consumption of these contaminated feeds by livestock has resulted in severe economic losses to the animal producer. In humans, ochratoxin has been implicated as the causative agent of a serious kidney disease called Balkan nephropathy. Ochratoxin A is a relatively stable toxin and can persist for months in nature. Concentrations in commercial corn in the U.S. have been reported to be as high as 110–150 parts per billion, and the highest concentration of ochratoxin A in contaminated feed has been reported as 27 parts per million. The acute lethal dietary concentration of ochratoxin in animals ranges from 2 parts per million in birds to 59 parts per million in mice. Younger animals are more susceptible to poisoning by ochratoxin A than are mature animals. Citrinin is a naturally occurring mycotoxin of wheat, rye, barley, oats, and peanuts that is often produced by several fungal species that also produce ochratoxin A, allowing for simultaneous contamination of feedstuffs.

Recently, much concern has been expressed about the effect of synthetic chemicals that have estrogen-like effects. We have touched on this lightly in previous chapters and will revisit it in Chapter 7. Zearalenone is an estrogenic mycotoxin produced by a number of species of *Fusarium* mold. It has caused a number of abnormalities in the reproductive system in animals and possibly humans. Corn grown in the midwestern states is the most common source of zearalenone in the United States. Warm summer temperatures in years with heavy autumn rains favor the growth of *F. roseum* in corn cobs. When these infected cobs are stored in open cribs, cold autumn and winter temperatures trigger the production of zearalenone. Pigs are the most common farm animals affected, often seriously. The health effects in pigs are indistinguishable from the effects due to large doses of estrogen. Recent epidemics of premature thelarche (development of the breasts before the age of 8 years) in children in Italy and Puerto Rico have raised speculation that zearalenone, or related residues in red meat and/or poultry, may have been responsible for the outbreaks. However, extensive testing of animal feeds and carcasses failed to substantiate this theory. The mycotoxin got into the animals through moldy feed. This, coupled with laboratory animal studies suggesting mutagenicity and carcinogenicity, make it prudent to avoid residues of these biologically active chemicals in the edible tissues of food-producing animals. It must be stressed that these are not synthetic chemicals but natural contaminants of foodstuffs.

There are a number of additional natural mycotoxins that are public health concerns, including the trichothecenes T-2 toxin, diacetoxyscirpenol (DAS), deoxynivalenol (vomitoxin), and nivalenol. Like zearalenone, the trichothecenes are produced by species of the genus *Fusarium*, but other fungal species are capable of producing up to some 40 chemically distinct entities. They have been implicated as the causative agents responsible for numerous diseases in domestic animals, as moldy feed is often fed to animals rather than humans. It has been associated in the former Soviet Union with a disease—alimentary toxic aleukia—in humans. The trichothecene mycotoxins, T-2 and nivalenol, were once implicated as the chemical warfare agents termed "yellow rain," allegedly used in Southeast Asia, although recent information appears to dispute this hypothesis. Chapter 10 discusses this nefarious use of both natural and synthetic chemicals in warfare and terrorist environments.

I decided to present the "gruesome details" of the actions of these natural chemicals because they have been responsible for a significant amount of disease and death in humans. *Nature is not inherently benevolent.* Humans are just one species fighting for survival on this planet. Since our modern technological civilization has not yet learned to control the weather, development of chemical fungicides have allowed food to be produced when Mother Nature was uncooperative. Although our modern agricultural system has improved means of harvest, storage, and transportation to help prevent fungal contamination, it is not perfect and contamination can still occur. When mycotoxins increase, so does the risk of serious health effects. This is especially troublesome and a public health hazard in developing countries. Some of the efforts of modern biotechnology are focused on generating seeds and plants that are inherently resistant to fungal infection. These will be discussed in a later chapter.

Our problem is that many natural constituents or natural contaminants of food are significantly more toxic to humans than are the pesticides the alarmists are so concerned about. As discussed in Chapter 4, Dr. Scheuplein of the FDA has publicly estimated that only 40 of the half-million annual cancer deaths in the United States in 1990 could be attributed to pesticide residues in food. In contrast, 38,000—almost a thousand times more—could be attributed to natural carcinogens. In another light, the average person consumes about 1.5 grams of natural pesticides daily compared to 0.1 milligrams of synthetic pesticides. This is a difference of 15,000-fold. If you are a coffee drinker, that one cup you are sipping on now while reading this book contains as many natural carcinogens, with an equivalent risk (or should I say nonrisk) of cancer, as in a year's consumption of produce containing trace levels of synthetic pesticides! Both the concentration and risk numbers are so low compared to natural background exposure that worrying about getting cancer from, for example, Alar® on an apple is equivalent to a baseball batter's concern about being hit by an angry fan's popcorn ball rather than by the Roger Clemens' fastball being pitched at him that second because the apple contains vastly more potent natural carcinogens. In reality, there is no risk from any of these events.

This discussion may cause alarm in some readers since there appears to be evidence that natural carcinogens found in our foodstuff is bad for us. First of all recall that Longfellow's *Reaper*, known today as cancer, strikes at increasingly old ages when the body's natural defenses begin to fail. Secondly, as discussed in Chapter 4, produce contains numerous anticarcinogens which counteract the sinister chemicals discussed above. This mechanism has to exist since all the evidence shows that increasing produce consumption is one of the best steps possible to promote good health and *avoid* cancer. Third, humans, and for that matter all mammals, have evolved over millions of years in this chemical-rich environment and have developed numerous protective systems to handle these chemicals. Many of the synthetic pesticides also resemble these natural compounds and are detoxified by the same enzyme systems already in place to cope with the natural ones. As discussed in Chapter 2, it is only when the dose is overwhelming, such as artificially occurs in laboratory animal studies created for regulatory purposes, that these systems are overloaded and toxicity ensues. When exposures are low, such as that seen with natural carcinogens in our foods, or are very, very low as seen with synthetic pesticide residues, there is no detectable harm.

Finally, defining a chemical's action and potential toxicity removed from its natural source is blatantly wrong. We must adapt a more integrated and holistic approach to risk assessment. Some of these concepts are embedded in current toxicology research aimed at defining methods for assessing risk of exposure to chemical mixtures as well as the emerging science of complexity. These approaches challenge the "reductionist" approach that is inherent to many aspects of modern science. Reductionism dissects a problem to its individual components, studies them in isolation of the whole, and then tries to infer behavior of the whole by summing up the individual results. This approach has tremendously increased our knowledge of underlying mechanisms at the molecular and genetic level but may not be appropriate to predict behavior at the individual or population level. When one moves up the ladder of biological complexity, variability and diversity occur, and models become incredibly complex. The whole is often more than simply the sum of its parts.

My research career has confirmed that the entire system must be studied if the behavior of the whole is to be predicted since looking at an individual component or single chemical can be misleading. The interactions become more important than the individual effects. A chemical is not working in isolation but rather is only one component of thousands. Although most people tend to be pessimistic and assume the worst case scenario when addressing mixtures, most of the times when one is exposed to multiple factors, the effects cancel out and true exaggerated responses, both good and bad, are the rare exception. This must be true since we have been exposed for decades to this "soup" of natural and synthetic pesticides, but as we saw in Chapter 4, we are actually living longer.

I must stress that it is not the science that is flawed. The flaws are the interpretation of the studies, the use of the results in the regulatory system, and the inability of the legal system to define laws capable of dealing with such a complex and constantly changing system. Isolated chemical studies are essential to defining mechanisms and increasing our understanding of diseases such as cancer. High-dose chronic animal testing in drug development may be useful to help us differentiate between different compounds so that the least toxic chemical is selected. Laboratory animal data is important, but its interpretation and implementation into regulations and legal statutes is being abused and taken completely out of context. Single-chemical, high-dose animal testing, or simple *in vitro* single-cell assays should not be used as qualitative check boxes to suggest that any level of exposure to a specific chemical means that there is a threat of cancer to humans. As Kenneth Foster and coauthors discuss in their excellent text *Phantom Risk*, a scientific test may be reliable and valid but *not relevant* to the problem at hand. Unfortunately, this often occurs in risk assessment.

These problems are more complex and are within the reach of modern science if science is allowed to be interpreted by its rules and not those of legal or historical precedence. We must adopt systems that rank relative risks and factor in known benefits. The "law of diminishing returns" will always operate, and it becomes more costly and ultimately impossible to reduce risk to zero. Zero risk to cancer is impossible because it is well below the inherent background risk of the natural components of food and the environment (ultraviolet or cosmic radiation). Maybe research efforts should be increasingly focused on defining why certain individuals are more susceptible to cancer or other adverse effects from any insult, synthetic or natural. Some goals of modern genomic research are specifically targeted to defining these individual characteristics that increase vulnerability to chemical insult.

I see an even greater threat looming on the horizon and that is the development of a legal strategy based on the psychology of exposure rather than objective toxicology or epidemiology. Exposure, not hazard or risk, becomes the critical criterion. In mainstream toxicology, *risk* requires *exposure* to some *hazard*. The new paradigm substitutes public "fear" of chronic poisoning as sufficient to claim damages against a chemical based on anxiety if the individual were only exposed to the chemical. This is clearly seen in the following quote from a recent court decision.

> That tort of negligent exposure to toxic substances gives rise to two completely distinct types of claims for compensatory damages: (1) Involving the increased risk of developing a disease in the future and (2) The other typified in the case before us involving the present injury of emotional distress engendered by the claimant's knowledge that he or she has ingested a harmful substance
>
> (Appellate Court of the State of California,
> *Potter v. Firestone Tire and Rubber Company* # SOI18831,
> December 29, 1993)

One does not have to prove that a chemical harmed a person, only that the person has a demonstrated fear of subsequent toxicity, even if none could be medically proven. Courts and juries have begun to rule on compassion. Exposure to a harmful substance *might* cause future problems. Proof is not necessary, only that the public at large is justified in having fear. Fear is evidenced by the prevailing public wisdom that a chemical is "dangerous." It is not science or expert testimony but rather the media and public opinion that would determine the safety of chemicals by influencing public opinion. What is truly frightening is that if this false fear becomes widespread, it may actually contribute to organic disease, as will be seen with Multiple Chemical Sensitivity syndrome in Chapter 7. One must work to dispel these unfounded fears.

We have demonstrated that there exists ample fear that low levels of synthetic pesticides may harm us. Only by using sophisticated state-of-the-art analytical chemistry tools can we demonstrate trace levels of exposure. However, there is *no* evidence that we have ever been harmed. As shown in this chapter, we are exposed to high levels of natural carcinogens in the produce we eat. We have exposure and hazard. We should be more afraid of this than of exposure to trace levels of pesticides. Yet we have amply demonstrated that eating produce is good for us, and thus most of these fears are unfounded.

However the *fear and anxiety* of harm from synthetic chemicals saturates our psyches, and thus truth and realities are ignored in favor of the tenets of *virtual toxicology*. As the legal case history develops in support of this argument, it effectively becomes law. Is this the proper forum for risk assessment? If hysteria and emotion are allowed to determine whether a chemical is toxic or not, society is doomed since the real hazards to our health, found in the natural world, are still there. I hope that this book can help dissuade this opinion.

The only thing we have to fear is fear itself.
(Franklin D. Roosevelt, First Inaugural Address, 1933)

6

Risk and Regulations

So far, I have focused on illustrating how safe our food is and how risk has been misrepresented, using actual data and scientific arguments. However, the process of risk assessment goes beyond science, as it implies that risks can be quantified, compared, and minimized, if judged to be unacceptable. This involves a subjective *value* judgment, which is often the root of most controversy. Scientific debate exists over the mathematical models use to extrapolate from high- to low-dose exposures and over the applicability of using toxic endpoints, which are relevant for one species or higher dose, to predict response in another species or lower dose. We are making rapid advances in our understanding of the mechanism of toxicity at the level of cells and genes. This, coupled with our increasing ability to integrate these into predictions of individual or even population responses, suggests that we are getting better at estimating risk. The controversy is in selecting what level of risk is acceptable and in determining if obvious benefits should be allowed to offset theoretical or real risk. When assessing risk, how do we determine that the level of exposure to the human population is hazardous? How do we set thresholds?

Workers studying the psychology and sociology of risk perception have discovered that the public is comfortable in comparing risks of similar origin, that is, food with food or drugs with the consequences of further disease. The easiest example, shown throughout history, is that most people are willing to put up with serious side effects if a drug will cure a life-threatening ailment. People with cancer will tolerate severe toxicity if life will be prolonged. Although great efforts have been made in reducing these effects, often they will still be tolerated if a life can be saved.

Similar risks should not be accepted, and are not, when chemicals are used as food additives. Everyone would agree that there should be no real risk to a food consumer from adding trace levels of pesticide or chemicals. However, one should realize that halting fruit and vegetable consumption is not the only answer since the enormous benefits covered in the previous chapter cannot be easily replaced. Although pesticide use may be minimized through wider adoption of integrated pest management strategies, some use of pesticides and herbicides will remain essential. Therefore, the goal is to set levels of pesticide exposure that are deemed *safe* for human consumption. Nowhere in the regulatory process to be discussed will any level of toleration or acceptance of adverse effects be permitted.

Based upon the data presented so far in this book, I hope you agree that the preponderance of evidence indicates that fruit and vegetable consumption is good for us. This is true even when natural carcinogens are consumed in the process. Since we have never been able to control the level of such natural toxins in our food, we must live with them, eat them, and we, paradoxically, actually live longer! However, with synthetic chemicals, we have developed risk assessment processes that generate even lower *theoretical* risks of cancer from trace levels of synthetic pesticides than from the natural food constituents. Essentially, using the same procedures previously discussed, the natural carcinogens in food are at levels above tolerance—at levels where our risk assessment techniques would predict adverse effects in humans consuming them. Yet, we know that this does not happen. Something is wrong.

We write regulations to ban synthetic chemicals without any evidence that our public health will benefit. It is not the science that is wrong, but rather the incorporation of selected aspects of this science into narrowly written regulations whose ultimate impact on public health is never assessed. Uncertainty and confusion are caused because we do not understand how to integrate molecular and cellular evidence of high-dose toxicity with the prediction of population responses. Laws that focus on regulating and biases against industrial man-made products both create hysteria when any toxicity test yields positive results, even though up to half of *all* of our natural foods would fail these *same* tests outright.

I will now take the approach of illustrating how such risks are calculated by using a more familiar threat of harm and even death. I will then apply the risk assessment process to establishing safe levels of exposure.

> How doth the busy little bee
>   Improve each shining hour,
>   And gather honey all the day
>   From every opening flower!
>
> (Isaac Watts, "Against Idleness")

My threat is being bitten by a poisonous insect or even spider. I will assume that a single bite is debilitating, or even deadly, and should be avoided wherever possible. This could be either the bite from the lethal Sydney funnel-web spider or from a brown recluse spider or a sting from a common bumblebee when the individual is allergic to it. Unlike some of the chemical risks discussed earlier, where the actual threat to humans is debatable, there is no doubt that venomous pests present a serious health hazard to certain individuals.

I will assume that the fictitious federal agency, the Venomous Insect Protection Administration (VIPA), is responsible for determining the minimal area where the risk of being bitten by an insect is remote. The first step in this process is to calculate the *safe area* where a human being and insect can reasonably be expected to cohabit. This will be experimentally determined by locking both a human subject and a devenomized insect in various size rooms to determine the minimal area where the person is not stung in a 24-hr. period. The reader should also keep in mind that in this process, only human subjects are being used, and thus there is no animal-to-human extrapolation necessary. Since there is no potential harm in these trials, college students are easily recruited for these 24-hr. visits as unlimited food is provided and a reasonable fee paid. It is determined from these trials that no bites occurred when the room was at least 144 sq. ft. (12-ft. x 12-ft. room). *The experimentally safe level of bee or spider (assumed similar) exposure is 1 bee per 144 sq. ft.*

Common to all regulatory agencies, this result now has a number of safety factors added into it, due to the uncertainties in extrapolating from experimental data to the real world. For the bee analogy, one could make arguments that maybe some individuals actually attract bees (something my wife always argues in favor of) or just that the anxiety of seeing and hearing bees must be avoided. Possibly the experimental bees used are relatively docile and not representative of wild types, such as the more aggressive South American "killer bees." In any event, there are a number of safety options available depending on the potential exposure, severity, and nature of the effect. We will be conservative and since the studies were done in humans, we will incorporate the minimal safety factor of 100. This could be justified as including a ten-fold safety factor for sensitive or bee-attracting individuals, and another ten-fold factor for protecting against "wild-type" bees. Applying this to our experimental data, the approved area for human cohabitation with poisonous insects will be 1 per 14,400 sq. ft. After converting units, the legal threshold for new human habitation of an area will be 1 insect or spider per one-third acre.

Some readers may be convinced that this approach is reasonable and would feel safe if this density were not exceeded. However, we should now change the perception of this bee sting and add the caveat that if a person is allergic to bees, an additional 10X safety factor should be added. This is precisely what is done if the toxic chemical being regulated causes birth defects or developmental problems rather than some other form of toxicity. The same holds if the target population is generally agreed to be more sensitive or generally exposed, for example, infants. This now puts our approved area for insect cohabitation with a bee-sensitive person (which is applied to all people) at 1 bee per 3.3 acres. It looks as if we are ready to eradicate the threat of bee stings from our existence.

If you are still with me and not getting paranoid that those 10 bees buzzing around your azalea bush in the backyard are eyeing you, let's apply the "zero-tolerance" Delaney clause to bee stings. Now, one would not be allowed to inhabit an area if a bee were even *detected*. One bite from a Sydney funnel-web spider will kill as will a bee sting to an allergic individual. Therefore, if an insect is detected, there is a real risk that a person may die. The reality of this hazard and desire to avoid such needless death is not debatable. If the VIPA were an existing regulatory agency, it would be reasonable for them to rule that to be absolutely safe, no one should be allowed to live in an area where bees are detected. This would promote the public health, as bee stings would be eliminated.

A bee sting is potentially at least as deadly and is even a more identifiable and avoidable threat than is a theoretical risk of getting cancer from eating a trace level of Alar® on an apple. The important concept to realize is that now your legal level of exposure is not tied to the actual risk of what the hazard will do to you but to the technology of detection of this hazard. Throw away the experimental data (1 insect per 144 sq. ft.) and even no-effect or safe levels (1 insect per one-third to 3.3 acres), and we see that VIPA now deems it danger-ous to your health if a single bee is ever detected. I will leave it to your imagi-nation on modern science's ability to detect insects. These types of laws, as illustrated in earlier chapters, now equate detection of a hazard to a calcula-ble risk. I will also not discuss the adverse effects of bee elimination, such as lack of pollination.

What is the real point of presenting this silly analogy? First there are numerous safeguards built into our regulatory system to compensate for sci-entific uncertainty and variability. These take a hard piece of experimental data and offer a safety cushion in case the trials were designed wrong or a sen-sitive population should be considered. This actual process, as applied to fruits and vegetables, is fully outlined in Appendix A. The numbers that result from this process are scientifically reasonable. However, many regulations now make the ultraconservative assumption that even one molecule of a chemical, which at lifetime exposures to high doses could induce cancer in a laboratory animal, might be hazardous based on detection rather than real risk. Any level is prohibited. The reader should recall the paradox of applying this to the natural constituents of food.

The subtle effect of changing endpoints by adopting detection rather than the safe level of a chemical is rarely communicated to the public, and thus exposure is equated to a real risk. In fact, a number of public panics have been started just by detection of a chemical without any assessment of risk!

As the bees are buzzing a few feet from my deck chair, I question why am I not attacked? Probably the bees are more interested in the nectar of flowers than in stinging me. This is where the *exposure* in the risk equation must be incorporated. This was previously presented as **risk = hazard** × **exposure**. The bee next to me remains just as *hazardous* as if it were sitting on my arm ready to sting. The difference is that with so many flowers growing in my yard and with the location of my jacket between the bee and my skin, I have additional protection drastically reducing my potential *exposure*. The bee and its hazard doesn't change, but the bee's access to me does, which lowers exposure and thus risk.

The same argument holds for exposure to a known carcinogen. Theoretically, one molecule of a carcinogen could mutate a specific piece of DNA (for example, a p53 tumor-suppression gene) in my body and cause a normal cell to become transformed to a cancerous cell. This cell could then undergo clonal expansion, evade detection by my immune system, and form a tumor. However, just as it is unlikely that the bee will sting me, it is equally unlikely that this single cancer-causing molecule will be absorbed into my body and cause an effect. Absorption by any route would require a high concentration of the molecules to be present, thus forcing the molecules to diffuse into the body. A single molecule cannot provide this driving force. It must then evade metabolism by my body's protective mechanisms and circulate through my blood and body until a susceptible cell is found. Using my bee analogy, the bee must first find and land on me before I slap it or spray it. One bee will not intimidate me, but a hive will. Similarly, although large doses of a chemical may cause harm, low doses may not. The analogies are endless. Crossing a street blindfolded may not be harmful on a deserted expressway at 5 a.m., but it is suicide on the same road during the 5 p.m. rush hour. This is the crux of the dose-response relationship. Also, if I am protected, it is unlikely harm will come. The same holds for chemical toxicity because our bodies have developed exquisite and very efficient protective mechanisms. Food, even with theoretically and experimentally proven natural carcinogens, is beneficial to our health. A *hazard* may always be present, but because of low *exposure, risk* is minimal.

Other lessons may be learned from this analogy, such as the difficulty of dealing with unusually sensitive or allergic individuals as well as environmental versus occupational exposure. The chances of bee stings dramatically increase if one is a beekeeper. Precautions are taken to ensure that such individuals are protected, as there is a real risk. The same holds for pesticides.

Applicators and manufacturers *are* exposed to high doses, so they must take precautions. Otherwise the serious toxicity predicted by animal studies may actually occur. One could argue that farmers are at an increased risk, however, (as discussed earlier) there is little epidemiological data to support this for chronic toxicity. Their risk is similar to applicators for acute toxicity. A person who is allergic to bees should not become a beekeeper. Does this mean, however, that beekeeping and flower gardens should be outlawed by the VIPA because some individuals who are allergic to bees may be stung?

> "Will you walk into my parlour?" said a Spider to a Fly;
> " 'Tis the prettiest little parlour that ever you did spy."
> (Mary Howitt, "The Spider and the Fly")

New laws are constantly being passed and our understanding of science increasing. There is no regulatory procedure in place to control risk to all classes of chemicals. Also, as is human nature, we tend to adapt existing laws and regulations to fit new circumstances even if the underlying concepts that resulted in the initial regulation are diametrically opposed to the new view. We apply regulations conceived for nuclear radiation or serious chemical toxicity to fundamentally different problems of low-level exposure to new generation pesticides or the inherently safer products of biotechnology. This misapplication of regulations, suitable for the original hazards, creates unnecessary concern for the new "threats" to which they are applied.

Confusion and chaos result when such paradigm shifts occur and old regulations are applied to new events. One such paradigm shift is the old view of a zero threshold for exposure to carcinogens. This was originally conceived using a radiation paradigm, which is incompatible with the evidence presented in Chapter 4, that a diet rich in natural carcinogens does not cause harm. Another paradigm shift is a result of the rapid increase in our knowledge of chemical carcinogenesis and molecular biology. This fundamental knowledge, coupled with parallel developments in our ability to quantitate these effects and form holistic and integrated models of chemical disposition and activity, allows us to dramatically improve risk predictions. However, no matter how sophisticated the predictions of exposure and hazard, the final step is a value judgment of how much risk is acceptable. That is an issue that I will not attempt to address. However, even without applying such sophisticated models, present levels of exposure to pesticide residues are apparently not causing us harm. Their restrictive tendencies may inadvertently be preventing some significant benefits of their use from being enjoyed.

Tolerance has been used as the field marker to suggest that exceeding it is dangerous to human health. The surveys quoted in Chapter 3 demonstrated that our produce is safe since the vast majority of it falls well below these tolerances. Yet alarmists sound the siren when produce is at or exceeds the tolerance. They ignore that enormous safety factors go into calculating this number (see Appendix A). They ignore the fact that daily consumption of chemicals at tolerance levels would be required to get at the estimated *no-effect* level for humans. Finally, they ignore the fact that *daily* and *lifetime* consumption of these specific "contaminated" products would be required to reach the *no-effect* level. If once a week you eat an apple, randomly contaminated at tolerance levels, that is one-seventh the risk of daily consumption, your concern should be that you are still at least *some 700 times below the safe level* predicted from animal studies. Should you worry about this?

Now I hope you see why I believe that it is criminal to sound an alarm when a pesticide residue is *detected* on produce at 100 to 1000 times *below* the tolerance. *The tolerance is related to a safety threshold, not a toxic one.* This misuse of data totally ignores the entire risk assessment process that is appropriately conservative and protective of human health. What is even more ironic is that the very chemicals that the public is made to fear in food at these below-tolerance levels are often used by the same individuals in household and garden pesticide products at much higher doses. These are generally proved to be safe for most individuals, making worrying about food concentrations millions of times lower a totally worthless experience that detracts us from more serious concerns.

My sole point is to make it absolutely clear that the *tolerance* used in the field to assess residue exposure actually relates to a *safe* or *no-effect* level, and not a level that means toxicity or disease will ensue. These are the tolerances discussed in Chapter 3 and used to determine the *violations* tabulated in Table 3.1. Secondly, there are numerous *safety* or *uncertainty* factors built into the determination of these tolerances that account for mistakes: the use of animals rather than humans or exposure to sensitive populations. Third, these tolerances mean *lifetime* consumption at these levels. Consuming a few apples at a detectable but subtolerance level should not be construed by any individual as being harmful! To state so is simply to lie!

 . . . out of this nettle, danger, we pluck this flower, safety.

(Shakespeare, *Henry IV,* I,II,3)

Chapter 7

Some Real Issues in Chemical Toxicology

The central tenet of this book is that consumption of produce containing trace or residue amounts of modern pesticides should not produce health effects because they are present at lower levels than those predicted to have *no effect* in animal studies. This statement is supported by the data showing the health *benefits* of consuming a diet rich in fruits and vegetables. Finally, we are all living longer and are not dying from a plague of chemical-induced cancer. I have no reservations in making the statement that the use of synthetic pesticides in modern agriculture benefits public health. The first part of this book illustrates both how modern science generates the data which can be used to make rational risk assessments, but it also clearly illustrates the paradox that often this data falls on deaf ears.

Many authors, including Howard (*Death of Common Sense*) and Murray, Schwartz, and Lichter (*It Ain't Necessarily So*) present many reasons why the public and press tend toward negative sensationalism when reporting scientific studies such as chemical-induced cancer. One common denominator is that such headlines sell newspapers, and that is the business they are in. Negative news is just not "newsworthy." In fact the same holds in the world of scientific publishing where negative data doesn't often warrant publication. However, this same science generates data which suggests that there are hazards out there which deserve attention, and more importantly resources to investigate. The American Council of Science and Health recently tabulated what they considered the 20 greatest health scares of recent times to differentiate unfounded health "scares" from proven health risks such as cigarette smoking. These included the cranberry scare of 1959, DDT in 1962, cyclamates in 1969, DES in beef in 1972, Red Dye # 2 in 1976, saccharin and hair dyes in 1977, Alar® in 1989, and cellular phones in 1993. Focusing on "non-issues" both detracts attention from real issues and makes it difficult to alert the public and press to give them the attention they deserve.

I am a toxicologist and fully aware that almost any substance when given at high enough doses may cause harm. Using high-dose toxicology data to describe the effects of consumption of residues in food is taking these data completely out of context and is wrong. Why do these issues continue to arise and detract our attention from the more pressing serious public health concerns?

Pesticide chemists should continue to make new chemical products with minimal adverse effects. They are needed to fight the ever-present insect vectors of viral diseases which do threaten our health. This philosophy has resulted in the modern development of safer pesticides and drugs. We should be concerned with minimizing occupational exposure and risk of adverse effects. We should continue our efforts to minimize environmental effects. However, we must realize that trade-offs may occur if we blindly pursue the unattainable goal of *zero risk*. High-yield agriculture may actually protect the environment by minimizing the need to convert natural habitats to farmland. Even with the major advances made in integrated pest-management techniques, which reduce pesticide use, some pesticides will continue to be used. These useful tools should not be summarily dismissed because we may be doing more harm than good.

Adverse effects associated with most agricultural pesticides occur at high doses that define the toxicity in laboratory animal studies. These effects may be seen with accidental or even deliberate poisoning, occupational usage, manufacturing, or environmental spills. The effects seen in high-dose laboratory animal studies and after high-dose human exposures are generally similar, allowing animal data to be used in some cases to study these effects. The rules behind making these extrapolations, based on the dose-response principle outlined in Chapter 2, are well known and form the backbone of modern toxicology. For many of the modern chemicals, such as the pyrethroids, even high-dose usage does not result in adverse effects. The newer pesticides that are currently used and being developed are significantly less active in mammals than those originally developed some 50 years ago. This attribute was part of the design process that led to their development. We cannot continue to drag the skeletons of past compounds out of their long-forgotten closets whenever the mention of the word *pesticide* is made! The newer insecticides and herbicides in use (see Tables 3.2 and 3.3) have less biological activity in humans and other animals, for they are more selective against either insect or plant pests. These new compounds are also less persistent in animals and the environment, thereby decreasing any potential for accumulation with resulting chronic toxicity. These differences in toxicity between chemical classes of pesticides can be appreciated by browsing through Appendix A. If toxic effects are detected, they are usually secondary to high doses that completely overwhelm the body's natural defenses.

What are the potential problems associated with pesticide usage when the issues of dose and newer chemicals are considered? Where should we focus our research direction, and how should we interpret these results and conclusions? There are a number of issues facing modern toxicology needing to be addressed which are directly related to the present discussion simply because they involve chemicals. However, the evidence which may be pertinent to ecological or occupational exposure, should not be used to argue against pesticide usage in modern agriculture relative to the endpoint of adversely affecting human health from exposure in food. The evidence is already "in" on the health benefits of produce consumption. Any new toxicology data generated must be interpreted in light of the experimental design with its inherent limitations.

A scientist can never prove that a chemical does not produce an effect. As discussed in Chapter 1, the language of science is statistics. We deal with probabilities. Experiments can show that it is highly likely that something will occur, but we can never prove without doubt that an event will happen. We also can never prove the converse—that an event will not happen—and this becomes our Achilles' heel when scientific data is used in the public arena. If an individual gets cancer and has been exposed to a chemical in their lifetime, that association can never be disproved because a probability can never be *equal* to zero. It can be very close to zero, for example 0.00000001—that is 1 in 100 million—however it cannot be zero. Anything, however unlikely or absurd, has a probability of occurring. *This lack of understanding of the language of science is the foundation of many of the chemical scares that presently occupy the media's attention.*

These issues associated with pesticides and other environmental chemicals which deserve some mention are the dilemmas associated with chemical mixtures, multiple chemical sensitivity, and the so-called "environmental estrogen" debate. They deal with organic chemicals and subtle effects which many believe to be associated with chemical exposure. In the sixties and seventies, it was our increasing ability to detect chemicals using advances in analytical chemistry that fueled our concerns about the safety of exposure to trace levels of pesticides. Today significant intellectual advances in many disciplines, coupled with the extremely sensitive techniques of molecular biology, genetics, and continued advances in analytical chemistry, raise concerns that chemicals may be causing us harm. The visibility of such chemical diseases is maintained in the public psyche with the release of movies such as *Erin Brockovich.*

Serious cases of chemical contamination do occur which may be related to human disease. However, all chemical exposure is not universally wrong.

The first issue to be discussed will be that of chemical mixtures. This area originally developed to address the issue of toxicity related to toxic chemical dumps and hazardous waste sites. We are dealing with the effects of relatively high doses of chemicals and are trying to find how multiple chemicals may affect the activity of other components in a mixture. The thrust of this research is to determine if toxicity studies conducted using single chemicals accurately predict the effects of multiple chemical exposure. This is a very valid and timely question. It remains a major problem in risk assessment since few chemicals are ever found alone in nature. As discussed in Chapter 5, food itself contains complex mixtures of natural chemicals whose toxicity has only been investigated on the basis of single chemicals. The same holds for chemical contaminants that make their way into food during the growing and finishing processes.

There are numerous mechanisms whereby one chemical may effect the activity of another, especially when all are present at biologically active doses. However, the results of such mixture studies must be discussed only in the context of the chemicals studied and of the experiments which were actually conducted. Often all that is reported is the simple conclusion that these studies prove that mixtures are different from single chemical exposures. Therefore, since mixtures of chemicals are found in our food, there must be a problem as toxicity studies have only been conducted with single chemicals. More chemicals must mean that even more toxicity may occur. This conclusion assumes that the toxic effects of all chemicals are *additive*, a hypothesis that actually is found rarely to hold in nature. The additional problem with this conclusion is that there is still no evidence of a real and significant health problem associated with eating food.

This research is significant when one considers all of the potential routes of exposure to chemicals in the home environment. For example, consider the exposure to pesticides that a person living in the average household receives. Exposure could include house treatments to protect against termite infestation, sprays to control ants and cockroaches, occasional uses of aerosols to "zap" flying insects, the application of many types of chemicals to control fleas on pets (in the carpet and in the lawn), and finally personal use of repellents when exposure to insects cannot be avoided. I would agree that such widespread use of pesticides should be controlled and more attention paid toward minimizing their use. These exposures are real and occur at levels that are known to have biological effects. United States regulatory agencies have realized this quandary of multiple exposures by a new regulatory approach termed "aggregate exposure." Allowable thresholds are beginning to take into account exposure at home, in the environment, and in food. This process, which has encountered some difficulties early in implementation, should prove beneficial in the long run. At the very least, estimates will be obtained of dominant exposure pathways.

Multiple exposure routes are a known clinical problem for pets when overzealous owners, in an attempt to get rid of fleas invading the household dog or cat, use pesticides in every niche of the pet's environment. This includes dipping and shampooing the dog, administering repeated sprayings, and using flea collars, aerosol pest strips in the sleeping quarters, flea bombs, and other products throughout the house. Lawns are often treated with even more chemicals. Not surprisingly, the pet may succumb to acute pesticide toxicity. The human inhabitants, especially children, may also exhibit some effects, although unlike the pet they can escape to other environments. Recent concern on pesticide use has arisen over their use in airlines to control insects from entering the United States on international flights. These exposures are real and have resulted in clinical toxicity in people. Faced with this level of household pesticide exposure, I would strongly argue that it is a waste of everyone's time and energy to be concerned with the additional exposure resulting from eating trace levels of pesticides in our produce.

Much of the research on chemical mixtures is involved with the above concerns, complicated by occasional exposure to prescription drugs and over-the-counter remedies. There are interactions which have been detected and should be addressed. Interactions can be classified as additive ($1+1=2$), synergistic ($1+1>2$), or antagonistic ($1+1<2$). Obviously, the direction of the interaction is important. Many interactions are due to the potential effects of chemicals on:

- Inducing increased liver enzyme activities
- Inducing mild organ toxicities which alter the subsequent response to other chemicals
- The competition for binding to selected molecular targets
- The ability of certain chemicals to "promote" the development of cancer

Antagonistic interactions also occur, reversing the direction of these effects, such as occurs with enzyme inhibition, etc. Concerns of interactions are critical in the area of pharmacy and medicine in the guise of drug interactions. Serious health effects occur when improper combinations of drugs are administered, such as occurred recently with dietary pills. However, all of these effects are detectable in laboratory animal studies and follow well-defined dose-response relationships. Like any other toxicological effect discussed in earlier chapters, a no-effect level is defined, and the safety of chemical exposure should then be interpreted relative to these levels which have appropriate safeguards built in.

Many of the initial insights in the field of mixtures and drug interactions were made relative to studying the health effects of smoking and alcohol consumption. Tobacco smoke is a complex mixture whose individual components do interact to produce adverse reactions. Recall from earlier discussions that the primary cause of death from cancer in the United States is related to cigarette smoking. Similarly, chronic alcohol consumption changes the enzyme profile of one's liver which then affects the way other drugs or substances are metabolized by the body. In addition to resulting in chronic liver disease and ultimately liver failure, the activity and toxicity of medicines may be adversely affected. This interaction is very important since many commonly used drugs are metabolized by the liver and thus susceptible to producing adverse effects, including further liver damage, when given to people with preexisting liver disease. This concern holds for the widely prescribed cholesterol-lowering "statin" drugs. It has recently become an issue with the common over-the-counter drug acetaminophen where liver toxicity was found when these analgesics were taken with alcohol.

Many other risk factors are only expressed in individuals who are chronic smokers or suffer from alcoholism. These are serious issues that must be addressed and may be relevant to certain drugs and chemicals. However, they are not likely to be relevant to the issue of isolated exposure to residue levels of different pesticides on food as would occur when one eats a diet of produce obtained from the local grocer. It is a mistake to confuse them.

Some mixture studies have used drinking water exposures to low doses of pesticides and demonstrated discernible effects in the cells of animals studied although gross disease was not detected. The problem with many of these studies is in the interpretation of a subtle change in the structure or function of a cell relative to the effect on the whole animal's well-being and longevity. In one study conducted by Kligerman, rats and mice were constantly exposed to 6 pesticides and a fertilizer in their only source of drinking water. Cellular defects in cytology (cytogenetic) were detected in rats and at the highest concentration in mice, but no other adverse effects were noted. Unfortunately, the potential health significance of the defects noted are not known and are hotly debated. These studies clearly indicate that chemical exposure occurred and the cytogenetic effects observed may be useful as so-called "biomarkers" of exposure. However, they do not evaluate the consequence of the presence of such biomarkers and how the body reacts to them. These biochemical effects are minimal and the exposure levels exceed what would occur with intermittent consumption of commercial produce containing pesticide residues at the levels documented in Chapter 3. It should be noted that these studies were not conducted to assess the effects of pesticide residues on food. These findings may be significant to groundwater contamination near agricultural or industrial sites, where a people's exposure would be constant should their sole source of drinking water be from a contaminated aquifer. They should not be extrapolated to conditions far different from the experiments conducted.

A similar concern occurs when very sensitive "in vitro" studies indicate a chemical effect. In vitro studies take a cell or component of a cell out of the organism and maintain it alive in the laboratory. The effects of chemical exposure to these cells are then studied. This is a very sensitive and useful technique in toxicology when properly interpreted. For example, one could study the effects of chemicals on cells of the nervous system and detect potential effects. However, these same cells in the animal are very well protected by their location from ever coming into contact with this chemical. Secondly, some responses produced in isolated cells would be immediately counteracted by other cells in the intact animal. These tests have their place in toxicology. They can be used to probe the mechanism of action of an effect seen in animals, or be used as a screening tool to select compounds for further study in intact animals.

There are numerous cases where the presence of one chemical may potentiate the effect of another. At high doses the pesticides are classic examples. Since most of these chemicals actually affect the same target site, combinations may be additive. Thus lower-level (*not trace level!*) exposure to mixtures may accumulate to the point where an effect is seen. This is a common occurrence with the organophosphate insecticides and is the source of the problem for our poor flea-infested dog. In this case, all organophosphates interact with the same molecular target—acetylcholinesterase enzyme at nerve synapses. Thus if two organophosphate insecticides are given at doses that individually would not elicit an effect, the combination may cause enough inhibition that an effect is seen. Subchronic exposure to relatively high levels of these pesticides may cause an inhibition of these enzymes, which are detectable in the laboratory or hospital, and could potentiate the toxicity of a subsequent dose.

This concept is the origin of the "water barrel" analogy, where one drop of water spills the barrel because the barrel was already on the brink of overflowing. The analogy (used by proponents of multiple chemical sensitivity and discussed later) says that after a lifetime exposure to "toxic" chemicals, that last single exposure is enough to finally cause disease. The interaction discussed above with organophosphate insecticides is not applicable since it cannot be generalized across all chemical classes. These documented effects are occurring at relatively high doses and are specific to organophosphates and interactions with their specific target enzymes. They are readily detectable in our standard laboratory studies. Although low doses of chemicals may also produce additive effects, the individual components may be eliminated before any effect is seen or protective mechanisms of the body intervene. In the case of organophosphates, the enzymes themselves may be eliminated from the body and liver enzymes would be induced to metabolize the excess chemical. The preponderance of accepted scientific data supporting the importance of interactions is at relatively high exposures or concentrations associated with classic and measurable endpoints of pharmacology or toxicology.

Another problem with mixtures is that many focus on the additive or synergistic effects seen. As discussed in earlier chapters, many times the effects are antagonistic and often decrease the toxic effect which would be observed if the chemical were administered alone. This is the situation seen with many antibiotic drugs where the action of one may actually prevent the second drug from exerting its effect. One antibiotic, a tetracycline, for example, may actually protect the invading bacteria from being killed by another, such as a penicillin. One must know the specific mechanism of action, dosages used, and the identity of the bacteria being treated to make this assessment. Many relatively toxic compounds that exert their effects by binding to important molecules in the body may bind to other chemicals in a mixture and essentially prevent themselves from exerting any effects on other molecules. This is the interaction seen when tetracycline is inactivated by calcium if taken together with dairy products. This lack of additive effects, and even antagonistic actions, may be part of the reason that our exposure to the host of natural toxins encountered in our food supply has not caused us any harm.

Our bodies have evolved to survive and even prosper in this chemical-rich environment. There are some thought-provoking discussions on such matters in the emerging literature of "Darwinian medicine," the new field of study that attempts to explain human reactions to various environmental stimuli in terms of evolution and natural selection. Our bodies today are a result of eons of evolution that resulted in the development of Stone Age hunter-gatherers whose diet was marked by diversity. We evolved in a "soup" of chemical toxins and developed efficient means to protect ourselves. One could even make an argument that exposure to such substances actually keeps our bodies' numerous defense systems "primed" and in operating function, ready to increase activity should a serious toxic threat occur. This is consistent with the concept of "hormesis" presented earlier. We do not know the effect of totally removing all natural chemicals—an impossibility anyway—from our diet. However, as the proponents of Darwinian medicine clearly stress, the pressure of natural selection only developed adaptations to keep us alive about a decade older than the age of reproduction, when the genes controlling these traits could be passed on. Beyond that, our bodies literally start to become dysfunctional. Cancer is one of the manifestations of this aging state termed "senescence." Our susceptibility to disease increases, and our ability to make repairs fails. We do not have defenses to keep us going much beyond age 85 and apparently reach a wall at 115. To quote Drs. Nesse and Williams in their book *Why We Get Sick: The New Science of Darwinian Medicine*:

The extraordinary low death rates of the last few centuries, and especially in the last few decades in Western societies, show that we live in times of unprecedented safety and prosperity.

Trace levels of mixtures of pesticides are not a problem when looked at in this context. There are other interesting perspectives if this line of thought is followed; however, I will return to these in the final chapter.

As discussed in many earlier chapters, it is often difficult and even dangerous (read as outright wrong) to take the results of a single and limited study and extrapolate its findings to different chemicals and exposure scenarios. Studies are usually conducted for very specific purposes and have limitations to their interpretations. This is a defined problem that has been addressed in the professional literature. It is a rare event when a basic scientific research study designed to probe the mechanism of a chemical interaction is conducted according to the exact design pertinent to a real-world field exposure. The basic research and regulatory community often have difficulty making basic research and real-world exposures interface.

At a certain level, this problem of overextrapolation also seems to be commonplace in the popular literature dealing with chemical mixtures. The scenario goes something like this. We know chemicals A and B are bad for us. At levels normally found in a certain environment, there apparently is no effect. However, results of studies on other chemical mixtures suggest that sometimes chemical effects are additive; therefore, since A and B are present in our foods, albeit at very low levels, we should be concerned and ban their use until we can prove they are safe. This is the depth of many of the arguments that are presented to the public today. They might be credible if the "mixture" studies quoted were actually conducted on chemicals A and B. Often they are totally unrelated substances that show additive effects. Since they could be classified as pesticides or are agricultural chemicals, the results are assumed applicable to all pesticides at any concentration, conclusions which are a blatant misuse of science.

There is even a simpler and potentially more damaging problem relating to semantics. Investigating the toxicology of chemical mixtures is becoming a major area of research. The findings are relevant only to the chemicals and doses studied and for the endpoints actually examined in the experiment. One should only extrapolate results and conclusions based on the data and not on the title of the research. This has become evident in popular writings on a problem which has recently been in the public's eye. This is "Multiple Chemical Sensitivity" (MCS) or so-called "Environmental Illness" or "Sick Building Syndrome." A more appropriate term which I believe better describes these syndromes is Idiopathic Environmental Intolerance (IEI), where idiopathic is a term used to indicate that the cause is not known. I caution the reader: just because the phrase "multiple chemical" appears in the term MCS, all of the scientific work done under the rubric of "chemical mixtures" is not automatically relevant. MCS is not synonymous with studying the toxicology of chemical mixtures, and the scientific findings of the latter were not meant to be blindly extrapolated to provide justification for a mechanism of MCS. The MCS/IEI syndrome must be discussed on its own merit.

Environmental illness and MCS are terms which are increasingly making their way into the media, popular press, legal system, and even insurance coverage guidelines. It is a syndrome characterized by extreme sensitivity to any and all levels of organic chemicals. Most synthetic pesticides are organic chemicals and thus are implicated as being responsible for MCS. Numerous other products fall under this umbrella, including solvents, fragrances, preservatives, many food additives, detergents, plastics, smoke, and a host of other modern chemicals. Pesticides are acknowledged by all as being only one potential component of this problem. Their complete elimination from the planet would not eradicate this disease even if they were proven to be partially responsible.

There have been a few other syndromes occasionally confused with MCS that should be briefly addressed. One is the Gulf War Syndrome, in which some American soldiers suffered a variety of illnesses after returning from that conflict in 1991. Recent studies suggest that if such a syndrome actually exists, it may have been caused either by multiple vaccinations or by exposure to drugs, insect repellents, and/or burning oil wells associated with this conflict. These were *not* low-level exposures and fall more under the umbrella of true chemical mixture toxicology discussed above. A Department of Defense analysis of 10,000 veterans and their families failed to find any evidence of a unique disease but instead attributed all symptoms to a panoply of common diseases unique to the soldiers only because of a temporal association with the war. Other soldiers in the control groups, who did not go to the Gulf, suffered similar problems; however, they had no singular event to attribute causation. In fact, one scientist, Professor Hyams, has noticed that veterans returning from wars throughout history displayed complex syndromes that could be confused with the modern Gulf War Syndrome. In the American Civil War it was "Soldier's Heart"; in the Boer War it was "Rheumatic Condition"; and in other conflicts it was "Shell-Shock" and "Post-traumatic Stress Disorder" (PTSD). Some of these postconflict illnesses were attributed to soldiers carrying heavy backpacks or having prolonged exposure to adverse weather conditions. There is no doubt that the extreme stress of combat would potentiate or aggravate many clinical conditions. In fact, some studies suggest that stress alone may potentiate the action of many drugs and chemicals and even increase the ease by which drugs may gain access to the normally protected central nervous system. This common problem of background disease and the trouble epidemiology has in assigning cause were fully discussed in Chapter 1 and is pertinent here.

MCS is one of the latest mystery diseases to confront the scientific and medical communities. The term was first coined by a physician, Dr. Theron Randolph, in 1962, spawning the "clinical ecology" movement. Symptoms may include headache, numbness, tingling, muscle weakness and pain, joint pain, fatigue, dizziness, confusion, inability to concentrate, slurred words, difficulty to find words, memory disorders, odor hypersensitivity, as well as other nasal and throat problems. In some cases skin rashes may be involved. This generalized deterioration of health occurs over months to years, often associated with a change in environment. One of the major problems is simply to define MCS or IEI precisely enough that detailed cause-and-effect studies may be conducted to gain insight into its management. The hallmark of this syndrome is the involvement of multiple organ systems reflected by this complex symptomology. As with many other syndromes on the fringes of "established" medicine for which a defined cause is not presently known, most professional attention is focused on attacking the outlandish theories of causation since they are not consistent with the accepted paradigms of medicine. However, patients are suffering from something, and it behooves the medical establishment to find a cause. The MCS syndrome is beginning to be studied in this light and hopefully the seed can be separated from the chaff.

One problem confronting the study of MCS is that it is rare, with estimates ranging from a high of 1% of the population to a percentage much lower than this if precise diagnostic criteria are used. This should be contrasted with the magnitude of common allergies in the population which are estimated to affect 25% of us. If there is a problem, what can be done? How does one weigh the benefits of helping less than 1% of the population when 99% may be adversely affected by the remedy? This is not a question of science but one of public policy, politics, and common sense. To address this dilemma, it is important that the facts be correct.

The first problem with MCS is defining the syndrome so that a patient population can be identified. I present the definition as published in a 1992 MCS Workshop:

> Chemical sensitivity is an abnormal state of health characterized by intensified and adverse responses to components found in food, air, water, or physical surroundings of the patient's environment. The provoked symptoms or signs may be chronic, relapsing, or multisystem and are triggered by levels of exposure that are generally well tolerated by most people. The diagnosis is suspected on the basis of history and physical examination, and the condition may be confirmed by removing the offending agents and rechallenging patients

under properly controlled conditions. The clearing or improvement of symptoms or signs by removing the suspected substances and the recurrence or worsening after specific, low-level challenge is highly indicative of environmental hypersensitivity and, in this context, chemical sensitivity.

From this definition, the reader should appreciate that this is a complex clinical syndrome that may be triggered by a number of different chemicals or physical environmental triggers. Although almost any sign has been associated with it, the symptoms often include dizziness, headaches, nausea, lack of concentration, weakness, and fatigue. A significant amount of increased public, legal, political, and medical attention has recently been focused on MCS because of the recognition of the "Sick Building Syndrome" in workers occupying modern airtight buildings. This may be a serious problem and is probably related to modern recirculating air-handling systems which increase exposure to chemicals, molds, or viruses found in a building. I urge the reader to remember that this book is primarily addressing chemical food safety and not the potential effects of *any or all* synthetic chemicals to which an individual may be exposed. It is feasible that higher level and chronic exposure to chemicals in an environment such as an airtight building or the battlefield environment could sensitize an individual to a chemical. If subsequent exposure occurs in food, then a reaction might occur.

MCS is most similar in presentation and possibly etiology (causation) to many allergies for which, once sensitized, an individual will exhibit an exaggerated response to chemical contact, no matter how low. Poison ivy is a case in point. For some MCS victims, classical techniques used by allergists to diagnose and treat these patients may be appropriate. In these cases, the patient is simply allergic to a synthetic rather than natural allergen. Many food allergies fall into this scenario. However, this is *not* the typical presentation for most MCS patients for whom the routine clinical tests are essentially normal. When immunotoxicologists addressed potential chemical causes of immune system disease, they saw that the low levels of organic chemicals postulated to cause MCS were negative in their very sensitive batteries of scientific tests. In other words, the chemicals which cause suppression of various components of the immune system in experimental studies only did so at much higher concentrations, similar to toxic levels, or, in some cases, were of a totally different chemical nature than those purported to be responsible for clinical MCS.

Medical attention has focused on the Sick Building Syndrome, where inhalational exposure to a wide variety of chemicals used in the construction and maintenance of modern buildings may be making some individuals sick. The one common denominator among these individuals (as well as in other cases of MCS) is *inhalational* exposure and a central role for our sense of *smell* in the clinical presentations. Pesticide exposure is often detected by smelling the solvents in which the chemicals are dissolved, curiously not the pesticides themselves. These volatile solvents are also found in many other products such as cleaning agents, dyes, newsprint ink, perfumes, fuels, and carpet adhesives. Another potential culprit may be "off-gasing" of formaldehyde from some building materials. In other cases, there may actually be acute pesticide exposure which is characterized by many of the same symptoms. These symptoms would be like those of acute organophosphate toxicity, similar to what our flea-infested dog experienced, and can be detected by higher concentrations of pesticide exposure. Again, this may be a real toxicological problem but is not pertinent to pesticide exposure secondary to eating produce.

There are a few hypotheses developing in the medical community about some potential causes of "true" MCS. This work is focused on psychological causes and the role of smell (olfaction) in triggering this syndrome since it is a common thread that weaves its way through MCS and IEI patients' histories. Some workers attribute up to 50 to 70% of MCS as being a psychological problem of mental association and conditioning, a cause amenable to psychiatric treatment. In these cases, at some early stage patients subconsciously associate signs of a disease with the smell of a volatile organic chemical. All that is necessary to trigger a subsequent episode is detection of the chemical. This then launches a cascade of physiological responses similar to that which occurs when an allergic individual is "reacquainted" with a particular allergen. In both cases, the chemical is the trigger, and our body then launches an exaggerated response. For classical allergy, the chemical detector and amplifier resides in the immune system. For MCS, the detector is our sense of smell and the amplifier is our brain and nervous system along with the rest of the body it controls. Susceptible individuals may have underlying psychological problems such as depression or stress, hormonal imbalances, compulsive behavior syndrome, genetic predispositions, or an inappropriate belief that chemicals cause harm. Whatever the cause, this syndrome produces a true and potentially debilitating disease.

In some cases relevant to food, our sense of taste may serve as the detector. Conversely, the patient may not be conscious of detection except by exhibiting the response. Early in our evolutionary history, this reflex probably served a very useful function since food which previously caused problems could be eliminated before absorption occurred. Since taste and smell are our body's most sensitive detectors, this reflex prevented unwanted intake of poisons. Our olfactory systems can detect chemicals at concentrations far below those necessary to induce a biological response. Today the mind is inappropriately associating certain distinctive organic smells with harm.

Patients with MCS believe that their symptoms result from low-level chemical exposure. Many patients respond to signs by avoiding situations of perceived exposure. This action reinforces the belief that the signs will abate when the individual believes there is no exposure. To some psychiatrists progressive withdrawal and the resulting disability associated with this isolation are the label of severe chemical sensitivity. The media has covered severe cases, such as people living in remote environments or sterilized and chemical-free quarantined homes. Even popular television shows, such as *Northern Exposure*, publicize the subject. The reader should differentiate this from individuals with nonfunctional immune systems who were once treated by living in sterile environments due to their inability to combat any infectious disease. Individuals with this type of MCS may be treated by psychiatric intervention, and some antipsychotic drugs have also been effective in breaking this vicious cycle. The association of psychological factors with MCS does *not* diminish the medical reality of the disease!

As the reader can appreciate, the above description of this syndrome is, by necessity, incomplete and simplifies the complex presentation of this disease. There has been some overlap with the signs of Chronic Fatigue Syndrome, which might have an infectious disease trigger. Both have psychological manifestations. What is curious is that some researchers have even reported evidence of this "hypersensitivity syndrome" being discussed in the seventeenth century. This obviously was not due to the products of twentieth-century chemistry.

To me, the frightening aspect of this psychological hypothesis is that weak associations could easily be reinforced by higher cognitive areas of the brain if the individual truly *believes*, based on media accounts, that low-level chemicals are uniformly harmful to us. In these individuals, perception of potential harm is the important factor, as opposed to toxicological activity of the chemical itself. This perception is translated into serious physical disease. The reader should recall our discussion of the placebo effect in Chapter 1 since many aspects of this phenomenon are pertinent. Continued media exposure of these issues reinforces this perception and actually creates a definable syndrome.

Most readers can appreciate that our sense of smell is extremely sensitive and is capable of detecting truly minute concentrations of volatile aromatic substances. It is no surprise that most perfumes are made from the same class of compounds. The scenario presented above is capable of explaining individual reactions to very low concentrations of chemicals in our environment. Some recent studies have even suggested that there may be a unique anatomical pathway in our nose that allows chemicals to directly enter the nerves responsible for smell. These olfactory nerves, with receptors in the nose, then connect to the limbic system of the brain (see below for a definition of the limbic system). This pathway has been clearly demonstrated with manganese transport after inhalational exposure. Note, however, that oral exposure of this metal does not result in access to a similar pathway to the brain. This olfactory nerve pathway provides a direct link between the outside environment and the nervous system. In some sensitive individuals, this provides a biological mechanism for the wide variety of symptoms associated with MCS and the allergic-like amplification process seen where minute exposure results in an exaggerated response. However, exposure to chemicals in food would not function by this mechanism.

In these cases, the amplification would be mediated through physiological rather than psychological actions of the brain. There are even specific mechanisms in the brain for chemical sensitization (termed *kindling*) to produce such exaggerated responses. At this time, this is a hypothesis and is not substantiated with experimental evidence. Amplification through the olfactory nerve might represent discovery of an entirely new type of reaction to chemicals, mediated by the olfactory and nervous systems rather than the immune system, which responds to allergens. Similar to the psychological mechanisms, learned behavioral patterns could then modulate this response since the limbic system is an evolutionarily primitive area of the brain that is known to control and influence emotions and basic physiology. The nerve connections are present for behavioral input into this response. If this olfactory/limbic model holds true, there should be clear avenues for drug intervention. However, there should be just as much concern that *false impressions* of low-level chemical toxicity could exacerbate otherwise manageable symptoms. In this light it is understandable that pets suffering from all the other types of diseases their human owners experience, such as cancer, allergies, infections, AIDS-like viruses, etc., do not appear to have MCS.

A very different picture is painted if one examines the popular writings on MCS. Low-level chemicals are blamed for everything from cancer to MCS. The problem this author has with these accounts is that they are anecdotal and make conclusions based on unsubstantiated associations, for example, between pesticides and the "cancer epidemic." As the reader of this book should now appreciate (since we are not in the midst of a cancer epidemic and eating veggies even with pesticide residues is apparently a healthy practice), it is difficult to objectively address arguments supported by this "pseudodata." The suggested remedies fly in the face of a substantial body of medical science whose validity has already been proven many times over as evidenced by our increasing life span and irradication of many diseases. People who have MCS have a syndrome that merits attention, but the playing field must be narrowed and defined.

Where does this lead us relative to the safety of residue levels of pesticides in food? Most of the focus on MCS has been related to inhalational exposure of pesticides and volatile organic chemicals. This is central to the olfactory nerve pathway for chemicals to reach the brain. This is not relevant to our problem with food unless a previously sensitized individual is "triggered" by detection of a chemical in food. Food allergies, such as that widely reported for peanuts, are a serious problem but are not unique to synthetic chemicals. As discussed earlier, I would judge that multiple exposures to pesticides in a closed environment, similar to that seen with our flea-infested dog, might produce signs of acute pesticide toxicity. However, these concentrations are significantly higher than what we are talking about in this book. If MCS exists and is related to pesticide exposure, it is conceivable that detection of the chemical in food could then trigger a response. Does protecting this segment of the population from all traces of organic chemicals serve the well-being of the vast majority? If society decided in the affirmative, then all perfumes and most other chemical products including disinfectants, plastics, and many medicines would have to be banned. It would seem to me that pesticide residues are the least of our concerns and should not be the first to be eliminated by MCS sufferers since they do benefit the large majority of the population. Instead, I would focus research on preventative strategies, be it drugs targeted to the olfactory/limbic connection, psychotherapy, or, conceivably, even biofeedback approaches to control the response.

The final issue with which I would like to close this chapter is the so-called "environmental estrogen debate." The focus of this debate is not on pesticide residues in vegetables but rather on chemicals in the environment with certain properties that may give them biological activities similar to the female sex hormone, estrogen. In the field, some workers believe that excessive use of certain classes of chemicals has resulted in signs of toxicity which mimic the actions of estrogens. This finding, relative to ecology, was then coupled with an observation that the sperm count of men has decreased over the past fifty years, resulting in a media alarm to the general public. A conclusion is then formed that chemicals must be the causative agent. Since some pesticides (especially many of the older chlorinated hydrocarbons) fall in this class, they have also been implicated. As we have seen for other toxic endpoints, the crusade against exposure to pesticides is then widened to include estrogen-like effects. This is entirely a separate issue from the focus of this book; however, it deserves some attention so that the readers can appreciate that these problems are distinct.

The primary observations are that some workers observed that there has been a trend since the 1940s of decreasing male sperm counts. Other epidemiologists have debated this finding since some studies have shown an increase with any potential decline only occurring before 1970. Most agree there has been an increase in sperm counts from 1970 to 1990, precisely when past chemical exposures should have had a maximum effect. Part of the problem of using sperm counts relates to why they were counted in the first place, which was generally as a diagnostic tool for infertility. Is it possible that so-called lower counts are an artifact of sampling? Regional differences also exist and, when factored in, eliminate any change in actual sperm counts. There is data that testicular cancer is increasing in young men, one possible cause of which *could* be related to environmental estrogen. A final observation is that some workers have suggested that the presence of similar chemicals may be correlated to breast cancer in women. Whatever the truth, let us assume that sperm counts have decreased, testicular cancer may be on the rise, and breast cancer is a serious disease. These are important problems that deserve study.

Another apparently related observation then enters the picture. Abnormalities have been observed in the sexual function or development of certain populations of toads, fish, alligators, and other wildlife. These may be correlated to high environmental concentrations of our old friend DDT as well as the environmental contaminants PCB and dioxins (see Appendix A for a description of these chemicals and why they are associated). Most of these are postulated to occur in areas associated with earlier spills or where acknowledged pollution is present. Recent field studies have actually implicated parasites as being responsible for many of these problems in aquatic animals. High-level exposure to chemicals in polluted waters may cause immunosuppression that allows parasite infestation. However, high levels of chemicals such as PCBs are judged bad by society and drive ecological reform. A good example is the ongoing cleanup of the Hudson River by General Electric to remove PCBs deposited there between 1947 and 1977. The food-related issue with PCBs in this environment is the consumption of fish that have bioaccumulated PCBs for their entire lives living in this environment. Levels may approach 30 ppm. Human consumption of such fish should be avoided.

Before we progress, let's look at the *data* we have. Sperm counts *may* be decreasing. Wildlife have been reported to have sexual problems in areas where spills or acknowledged pollution of some very specific compounds occur. There has been a suggestion that pesticides may cause breast cancer. The press reports these studies as proving that pesticides have been decreasing sperm counts in men, causing breast cancer in women, and playing havoc with the environment. The thread of fact which weaves through this scenario is that chlorinated hydrocarbons such as DDT, PCBs, and dioxin somewhat resemble the estrogen hormone and at *high* concentrations may have some estrogen-like action. Thus it is conceivable that environmental spills, or gross pollution, may affect the sexual and reproductive function of wildlife in these areas. But estrogenic effects in animals are notorious for being very species specific, that is, not being able to be easily extrapolated across different animals the way some other toxicological effects are. Whatever the cause, this is an ecological problem and may not directly affect human public health!

One study, judged almost definitive by many epidemiologists, specifically looked for a correlation between DDT and PCB compounds in women with breast cancer. They found none. The epidemiology community has accepted this lack of effect. Men naturally have circulating estrogen levels that must be extremely elevated to decrease male reproductive function. For a chemical such as a pesticide to actually exert estrogen effects, it must be present at very high concentrations. Let's face it, all women have circulating estrogen. It is not an inherently *evil* molecule. As with all other hormones, it is only when imbalances occur that trouble is seen.

What is the truth? Regardless of whether sperm counts are actually decreasing, there is *no evidence* that it is related to pesticide levels. Because some chlorinated chemicals such as DDT or PCBs could have this activity, this correlation was suggested, but there has been *no* data to support it. Many events have occurred on this planet over the past 50 years. Pesticide usage is just one of many.

The awareness of dioxin's potential adverse effects, and its reputation as being one of the most potent synthetic toxicants, was partially responsible for the magnitude of the dioxin food contamination crisis in Belgium in the late 1990s. Animal feed was contaminated with dioxin and worked its way throughout the livestock industry of the country. Dairy products had detectable levels present, which resulted in the condemnation of all foods made from dioxin-tainted products. This included, among other things, Belgium chocolates and products using butter, such as the world-famous Belgium waffle. The country's agricultural industries became isolated, exports stopped, and tourism vanished. The economic impact was so devastating that a new government was elected. This was all because of the presence of a chemical contaminant in the food.

Finally, one of the major concerns of this book is that all of these debates are based on a small class of chlorinated chemicals of which a few are pesticides and have been removed from the market long ago. *No one is implicating the bulk of pesticides that are actually used in agriculture today.*

I do not want to belittle the importance of this debate and its possible ecological manifestations. It must be stressed that scientists are not suggesting that pesticides chemically unrelated to chlorinated hydrocarbons would ever have this activity, yet all media reports use the generic term "pesticides" in their titles. Even for the pesticides that could have this action, such as DDT, it may only occur at high doses. The reader should not be concerned that eating produce with a minute amount of pesticides would ever produce these effects. I believe this is an ecological issue that should not be confused with the topic of this book.

In conclusion, I hope the reader appreciates that, despite the complete lack of evidence that trace levels of pesticides in produce are harmful to the consumer, the issue will never die. Whenever a new toxicity is described or a new syndrome remotely related to pesticide use is reported, the alarm is sounded, the anxiety level in the public increases and hysteria sets in. Regardless of how low they are, whenever pesticide levels are detected in produce, baby food, or some other product by a consumer or environmental group, the litany of potential pesticide toxicities, ranging from cancer to estrogen problems, is sounded, making people afraid and confused. Will this hysteria ever end? We are at the dawn of a new millennium. Let us use science to continue improving our lives and not allow hysteria to ban effective tools that have visibly aided public health, lest we revisit our earlier preindustrial existence marked by famine and truly devastating plagues.

> A truth that's told with bad intent
> Beats all the Lies you can invent.
> > (William Blake, "Auguries of Innocence")

8

Milk Is Good for You

The previous chapters have discussed examples of detecting the presence of certain classes of chemicals in foods. These chemicals, under some laboratory conditions, were reported to have produced adverse biological effects in animals. We were dealing with chemicals that at high doses or prolonged exposures were actually capable of producing toxicity. The problem of assessing their safety in food was one of establishing relative risks because of the extremely low levels of exposure. We were introduced to a class of compounds, the pesticides, which had dangerous relatives such as DDT in their "family tree." Even if the new generation of pesticides was significantly better "behaved" because of lower usage rates, decreased environmental persistence, and reduced toxicity to man, their family history was always being dredged up to haunt them!

Let us change perspective and leave the arena of the organic chemical pesticides. Instead, let us look at a recent example of the introduction of a different type of drug into our food supply. This drug, by every scientific and medical standard imaginable, is absolutely safe to humans consuming it. In fact, there is so much species selectivity in its actions that the compound is not active in humans even when it is injected directly into the bloodstream. In contrast to some pesticides, it does not produce adverse effects in the laboratory. Since this drug is a member of an entirely new class of compounds, it shouldn't have a "bad family reputation" to haunt it.

However, the development of this veterinary drug illustrates one of the clearest examples of public misconception and hysteria thrust upon society in recent years. This event was the November 1993 introduction of Monsanto

Company's recombinant bovine somatotropin (BST) into the veterinary food animal market. BST, also known as bovine growth hormone, rbST, rbGH, sometribove, and Posilac®, is a drug designed to increase milk production in dairy cows.

There is no *medical* or *scientific* basis for human health concerns when consuming milk from BST-treated cows. Normally, new drugs intended for use in food-producing animals are approved by the FDA without a whimper from the national news media. But BST was different, and it continues to generate a storm of protest. Some states have even passed laws challenging the FDA's authority to label and regulate drugs.

Why is BST so special? BST is a genetically engineered peptide produced by bacteria using recombinant DNA technology. Thus it became the first product of biotechnology to reach the American table. Its absolute safety is best documented in a quote from a study conducted by the Office of Management and Budget (OMB), of the executive branch of the federal government in January 1994.

> There is no evidence that BST poses a health threat to humans or animals. It has been studied more than any other animal drug, and has been found safe by FDA and many other scientific bodies in the U.S., Europe, and around the world. FDA also concludes there is no legal basis requiring the labeling of BST milk, since the milk is indistinguishable from non-BST milk.

This extraordinary investigation was conducted after FDA's own extensive review of the safety data concluded *no* remote risk to human health. As I will discuss later, the only factual debate which ever occurred was over the possible increased occurrence of an udder inflammation in cows receiving BST, which may be seen in any high-producing milk cow. Hindsight in 2002 suggests that this problem never materialized into a true clinical problem distinct from any high-producing milk cow. There was never evidence of any human health threat from BST consumption. Similarly, independent scientific bodies were convened in special sessions to review BST. All concluded that there was no risk to human health based on an analysis of over 100 scientific studies. Agencies giving their stamps of approval include the National Institutes of Health, the Department of Health and Human Services' Office of the Inspector General, and the drug regulatory agencies of Canada, the United Kingdom, and the European Union. Even the Congressional Office of Technology Assessment, the antithesis of the executive branch's OMB (which issued the preceding quote), agreed to its safety. *This drug and the milk produced from cows treated with it are safe.* Yet, if one reads the popular press or watches news on television, this is not the impression conveyed.

Compared to the issues of pesticide hazards (which we already discussed) and the hazards related to other products of biotechnology (which will be presented in the next chapter), BST has undergone the most scientific scrutiny of any drug ever used in food-producing animals. It has all of the properties of an *ideal* animal production drug. However, if one believes the arguments voiced by those opposed to the use of this breakthrough product, one would be convinced that consuming milk with BST would produce a variety of serious health problems.

These human health arguments usually assume abnormally high levels of BST are present in milk of treated cows. Yet no study could find a difference between normal milk and milk from BST-treated cows, even when using the same supersensitive techniques of modern analytical chemistry which caused all of the hysteria with pesticides.

Unlike the safety question surrounding pesticides, with their more complex chemistry and toxicology, the primary argument for the absolute safety of BST can be understood by anyone who has a friend or relative suffering from diabetes requiring insulin therapy. Today, insulin is manufactured as a recombinant peptide hormone similar in many respects to BST. Yet it is still primarily delivered the way it was decades ago, by hypodermic needles, since insulin cannot be absorbed after oral administration. Even if high levels of BST were found in milk (which they are not), it is impossible for humans, or for that matter even the cow itself, to orally absorb BST or peptides of any kind into the bloodstream and experience any biological effects.

Peptides are small proteins composed of amino acids. Most are naturally occurring hormones found in all animals that control basic biological functions, such as growth or blood glucose concentrations. Because of this need to control functions in response to rapidly changing conditions, animals have evolved molecules that have very short duration of action in the body. Since peptides are inherently fragile and unstable molecules, they are ideal for this purpose. However, the very properties that make them ideal messengers also ensure that they will not have effects in other animals consuming them. They are not persistent chemicals like the chlorinated hydrocarbons discussed earlier. They have very short half-lives and do not accumulate.

As far as your stomach and intestines are concerned, peptide drugs like BST and insulin are just tiny "protein snacks" which are very effectively *digested* into nutritious amino acids. If these drugs could be absorbed intact, then people suffering from diabetes would be eating insulin *pills* rather than using or receiving painful hypodermic needle injections.

Furthermore, the hundreds of thousands of children afflicted with growth deficiencies would also be taking pills rather than daily hypodermic *injections* of growth hormone. The cows receiving BST are given it by injection, an expensive and impractical way to give drugs to dairy animals. This area of research is very close to my own area of scientific expertise. My colleagues and

I have spent at least 8 years, and millions of dollars of research funds, working on ways to deliver insulin, growth hormone, and other peptide drugs into the body *without* using a syringe. Any pharmaceutical company that could accomplish this feat would capture a multibillion-dollar market. There is a clear financial incentive to produce a peptide pill. However, the conclusions from this research, which are published throughout the world, remain to this day that peptides are destroyed when they enter the hostile environment of the human digestive tract. They just become another protein meal. Their constituent amino acids are absorbed for a nutritional benefit just like any other food source.

Only with very modern and sophisticated techniques of making advanced polymer capsules can one ever hope to get even a tiny amount of intact peptide into the body. However, even with this emerging technology, only very small peptide hormones can even produce analytically detectable levels in the bloodstream, and these are not biologically active. These include thyrotropin releasing factor (TRF) and luteinizing hormone releasing hormone (LHRH), two peptides which consist of less than 10 amino acids each. Large hormones like BST with 191 amino acids or insulin with 51 amino acids are not absorbed using the most sophisticated technology after oral administration.

The result of most of this work is that drug companies have abandoned making peptide pills, and other novel routes of administration are being developed. These include inhalers and skin patches or transdermal drug delivery systems, the latter being my area of interest. This abandonment of peptide pills is especially noteworthy because the development of oral dosage forms for medications are viewed as the widely accepted and preferred way to give drugs to human patients. Also, oral dosing is by far the most profitable. This inability to develop oral peptide delivery systems for insulin and growth hormone is a current failure of the pharmaceutical industry and a major stumbling block for the widespread introduction of whole generations of therapies emerging from biotechnology companies. You can be assured that if human or bovine growth hormone could be absorbed after oral administration, it would have been done by now. Thus the simple fact that peptide pills have never been developed is absolute proof that there is no real risk of ever absorbing any BST from the milk of treated cows, even if it were present in very high concentrations, which it isn't.

Finally, the most convincing evidence in support of the absolute safety of BST relates to attempts to develop natural bovine growth hormone as a human drug. Natural BST, extracted from cows, was investigated in the 1950s as a possible injectable treatment for human dwarfism. Bovine insulin used to be the source of human insulin until biotechnology provided a better alternative. Both bovine and human growth hormone have 191 amino acids; however, 35% of the amino acids are different between the two species. *After inject-*

ing BST into the blood stream of human volunteers, it had NO biological activity and just plain didn't work. This conclusion was very recently reconfirmed using more sophisticated techniques. The bottom line is: BST does not work in humans.

Another scientific fact, which makes even the above arguments pointless and further confirms the absence of any risk to humans from BST, is how milk is processed. Pasteurization of milk destroys 90% of BST activity because of the inherent instability of a peptide. Thus even if one does not believe any of the above arguments, this simple fact proves the safety of BST-treated milk.

The only potential human health concern related to cows treated with BST was the possible increase in the level of a second peptide hormone related to the action of growth hormone, insulin-like growth factor-I (IGF-I). IGF-I is a 70 amino acid peptide and, in many respects, is structurally similar to insulin. When the milk from BST-treated cows was assayed for IGF-I, it had concentrations comparable to that found in human breast milk and well within normal levels of biological exposure. Since it is a large peptide similar to insulin and BST, IGF-I also is not absorbed after oral administration for all the reasons discussed above. However, there was still concern regarding its presence in infant formula because of the increased exposure and vulnerability of babies. Further studies showed that any IGF-I present in milk from normal or BST-treated cows was destroyed in the formula manufacturing process. Again the safety of BST-treated milk was demonstrated!

It is important that the reader understand the true significance of these findings. They clearly illustrate how easily the public can be frightened into illogical actions when there is absolutely no threat. From a human safety perspective:

- BST is destroyed in the process of milk pasteurization
- BST and IGF-I are both incapable of being absorbed into the human body after oral ingestion
- BST is inactive in humans when given by intravenous injection.

Therefore, based on all of the scientific evidence generated from both sides of the argument, one must conclude that there is no risk to human health of consuming milk from a BST-treated cow.

The main reason for the controversy over the introduction of BST was because of social and political factors which are beyond the scope of this book. The argument goes something like this. BST is acknowledged to significantly increase milk production. Increased milk supply would result in lower milk prices. This increased competition would drive less efficient, often smaller, dairies out of the marketplace because they couldn't compete with more efficient, and often larger, operations. Additional economic arguments have been presented that suggest this "glut" of milk would increase the cost of the

USDA's dairy price support programs and thus cost the taxpayer more. Market surveys also suggest that the public's fear of this product might decrease milk consumption, leading to an increased supply of milk due to a decreased demand.

The previously quoted OMB study refuted all of these arguments since the higher price support payments would be countered by reduced school lunch program costs to the government. These economic and political arguments must be debated on their own merits, not with false science to argue political points.

These concerns prompted grassroot protests, which even continue today, against usage of BST. However, it was the supposed threat to human public health that was the reason given for the proposed banning of BST. When people debated this point, the next "scientific" argument was based on the alleged increased incidence of udder inflammation (mastitis) in BST-treated cows. This would cause the farmer to use antibiotics, and we would then be poisoned by chemicals again! Also it was implied that BST must, therefore, be inhumane to the treated cows because of the discomfort experienced when afflicted by mastitis.

The reader must take note that the argument has now shifted from BST causing a human health effect to BST causing a "supposed" threat of increased antibiotic exposure because BST-treated cows needed to receive antibiotics for mastitis. There is not a shred of evidence that this has or ever would occur. It is time that we base decisions on data and not wild speculation.

There are a number of major weaknesses to these so-called scientific arguments. There is minimal human health risk associated with drinking milk of any source. BST itself does *not* increase the incidence of mastitis. BST increases milk production. Cows with high milk production *may* have more mastitis much as marathon runners *may* have more foot problems than the average sedentary person. In fact the FDA public Advisory Committee studying this very question concluded that other factors, such as season of the year, age of the cows, and herd-to-herd differences, were much more important in predicting whether dairies encountered mastitis problems in their cows. Should we outlaw milking cows in certain seasons of the year because cows are more efficient and thus more prone to mastitis, and then *possibly* prone to increased antibiotic use which *theoretically* could cause residues? It is the consensus of veterinarians and animal scientists that if the herd is properly managed from a nutritional and veterinary perspective, such problems can be minimized. This includes both seasonal fluctuations and management with BST.

The advent of modern herd health management programs, largely absent from the dairy industry even 10 to 15 years ago, are designed to improve dairy cow health through proper housing, nutrition, and preventive medical attention. The goals and objectives are similar to the IPM concept which reduces the need for pesticides. Simply put, healthy animals are the most profitable and least likely to require drug therapy. Like the pesticide programs, these strategies strive to decrease overall drug use based upon economic incentives. What about antibiotic residues? This subject is also very close to my area of professional expertise because of my involvement with the Food Animal Residue Avoidance Databank. Milk is one of the most tested and closely monitored food products produced in this country. There are multiple levels of safeguards provided to ensure that marketed milk, and also meats, do not contain violative drug residues. Even before recombinant BST was even invented, milk had been under close scrutiny because of its significance in infant diets. Also many people in our country are allergic to penicillin and can have severe reactions if any level of penicillin is present in milk. A similar situation holds for all antibiotics relative to the manufacture of cheeses and yogurts since any drug would kill the microbial cultures used to produce these dairy products. These economic and regulatory pressures assure that antibiotics are not present in milk at levels exceeding those considered safe by the FDA. *There is no residue problem in milk.*

The quality assurance programs instituted by the FDA and the dairy industry test *all* milk produced for the presence of antibiotics. Once again, very sophisticated, and often too sensitive, tests for drug activity are used. When a farmer or veterinarian treats a cow, a specific drug dosage is used and the milk must be discarded for a certain number of days depending on the product used. Records are kept on an individual cow basis. This process then ensures that the milk of cows is free from violative antibiotic residues before being collected in tank trucks that combine the milk from a number of dairy farms and deliver it to a dairy for processing. This detailed collection process helps to protect consumers even if mistakes happen to let a treated cow's milk get through. In this case, the tainted milk would be mixed and diluted with the milk from hundreds of nontreated cows from many dairies. This bulk tank is then retested for antibiotic residues, and the entire tank of milk is discarded if there is a positive drug test. For this reason, neighboring dairies contributing to this same bulk tank also have a strong influence on the quality of the milk allowed in the tank and an equally strong incentive not to be responsible for violations.

Today's milk processors demand milk free of antibiotics to meet consumer demands of food safety and avoid potential litigation involving product liability. As a final safeguard, the FDA also spot-checks milk and institutes legal action if a violation is found. This costs money to the offending producer and thus is another incentive to ensure clean milk. In fact, the FDA monitors milk for the same pesticide residues as they do for produce (see Chapter 3 for discussion and sources). Some studies indicate that milk is the source of trace levels of the fat-soluble pesticides that persist in our environment and thus may be exposing us to these added risks. We are again talking about trace levels below violative thresholds that have a miniscule risk associated with them. Since many of them are fat soluble, a simple way to avoid them is to drink skimmed milk; because there is no fat, there are no residues. This also lowers fat and cholesterol intake. However, remember that fat-soluble vitamins, such as vitamin A and D, are in milk, as well as calcium and a balanced source of animal protein—all factors which keep us healthy and protect us from cancer. The argument is the same as for pesticide residues on vegetables—avoid a theoretical risk of disease from residues and lose documented nutritional benefits.

Back to BST. If mastitis occurred in BST-treated cows and if antibiotics were used, the reader can now appreciate that there are numerous safeguards to ensure that milk is safe for consumption. But the biggest irony of this argument is why BST was developed. The "old" way to increase the milk production in cows was by administering low levels of antibiotics in a cow's feed. This helped decrease low-grade infections and improved the efficiency of "feed conversion." Most antibiotics are organic chemicals which can be absorbed if given orally. Before BST, the dairy industry was extensively attacked for this antibiotic use in cows because the resulting milk could have had residues that were harmful to the humans drinking it. It should be noted that all of these drugs are relatively safe to humans since they are routinely used to treat infections. As many readers can personally attest, to be effective the dose of an antibiotic must often be increased. It is unlikely that the minute amounts of drug that could evade the extensive testing programs and end up in milk could ever have an effect in the human body.

The real argument relative to antibiotic safety, which has some truth to it, is that chronically treating cows with low-level antibiotics may select for drug-resistant bacteria. This is a problem because this resistance could be passed on to bacteria living in humans if humans ate contaminated meat. These resistant bacteria could then cause serious disease. Numerous National Academy of Sciences' expert panels were convened to find evidence for this so that antibiotics could be banned from agriculture. However, such hard evidence, which would permit regulatory action, was never found. The only concern in dairy cows was the potential tissue residues in meat from culled dairy cows and calves. This problem has been addressed. Culled animals are animals which either have serious diseases and/or receive antibiotics to treat these diseases. They are *removed* from the dairy production herd, and their milk is never collected into the milk processor's tank or sold. The paradox is that BST was developed to eliminate this low-level use of antibiotics and the risk of contributing to bacterial resistance by improving milk production more naturally.

Despite all of this hard scientific evidence to the contrary, the unsubstantiated fear remains. Antibiotics are portrayed as capable of causing disease and are touted as "unnatural" organic chemicals. The upshot of this exposure was a massive increase in regulatory surveillance that resulted in the quality assurance programs described above. However, remember that this whole line of logic is irrelevant since it assumes that BST causes mastitis in a small percentage of treated cows and that antibiotics are then used to treat the mastitis. This doesn't occur. The debate has thus been shifted away from a BST effect to that of antibiotics. This is an excellent debating tactic but is misleading to the unprepared and should be unethical in the public health arena.

An ultimate irony of this whole story is that BST was partially developed to provide a drug naturally found in the cow's body that would gently increase milk production without providing any unnatural chemical residue in the milk of the treated animal. Monsanto was not alone in this quest since BST products were also being developed by other companies. The promise of biotechnology was to eliminate the need to routinely use drugs, such as antibiotics, and, use a product instead that nature itself produces to increase milk production. It is now being attacked because a *very small percentage* of cows treated with BST *may* require antibiotic therapy. There is a *remote* chance that some antibiotic may make it through the quality assurance gauntlet, survive massive dilution in a milk tank truck, and maybe make it into one glass of milk at a concentration that is, at most, 10 to 100,000 times less than a single antibiotic pill used to treat an infection. Should any rational person really be concerned with this?

The logic of attacking BST for fear of increasing antibiotic residues is flawed. The previously cited OMB study even reported on a favorable environmental impact statement with BST use. This is based on the fact that BST-induced increase in milk production per cow would decrease the total number of cows in our country. This could result in less pollution through decreased use of fertilizers, less cow manure, and less methane production. Why don't environmentalists quote these results?

Biotechnology is our best hope to decrease drug use in food-producing animals while increasing the value and quality of the milk itself. If anything, total use of antibiotics in cows might decrease if BST were used because of the closer management required to use the drug efficiently. As clearly shown above, there are no residues of either BST or other peptides. Even if they were present in milk, they would be inactive. Why is there such unfounded concern for BST? The only possible answer must be because of how it is manufactured. It is a product of genetic engineering produced with recombinant DNA technology. Thus it is marked for attack.

It must be noted that many drugs today, including insulin, the heart attack drug Tissue Plasminogen Activator (TPA), and many other biotechnology-based drugs, are produced using the same recombinant DNA technology. Hundreds are in the FDA approval pipelines. This supports the arguments in previous chapters on pesticides that people have different thought patterns when thinking about food versus therapeutic drugs. However, from a scientific and especially toxicologic perspective, they are identical.

Fear of the term biotechnology and recombinant DNA may be the driving force behind these concerns. Once the public becomes alerted to a possible threat to our food safety, our minds are primed for an adverse effect. In this case it is not DDT that is "the skeleton in the family's closet," but rather the threat of antibiotic residues. This "threat" is coupled to BST to give "scientific credibility" to the debate. BST is an excellent case study of how a drug, scientifically proven to be safe, can be attacked by innuendo in the absence of any supportive data. Unlike some pesticides, there is not even preexisting data to incriminate it—there are no skeletons in the closet. Yet a significant number of our population actually fears drinking milk from BST cows. This is a major problem that should be challenged. Failure to do so could have dire consequences on our future ability to substitute the inherently safer products of biotechnology for the organic chemicals typified by pesticides and antibiotics.

This entire attack is especially sad since drugs, such as BST, were created to eliminate the use of organic chemicals and to guarantee that milk would remain one of our safest sources of nutrition, suitable for even the youngest and most vulnerable among us, our babies. All evidence supports the safety and nutritional value of milk. Should this irrational fear of new technology drive our regulatory decisions? I think not. It definitely should not prevent one from drinking a glass of milk!

> Where ignorance is bliss,
> 'Tis folly to be wise.
>
> (Thomas Gray, "On a Distant Prospect of Eaton College")

Biotechnology and Genetically Modified Foods: "The Invasion of the Killer Corn" or "New Veggies to the Rescue"

The previous chapter demonstrated that some of the toxicological concern that is commonly directed to organic chemicals has been transferred to the relatively harmless peptide products. These compounds were specifically designed to be biologically "friendly" and have minimal adverse side effects. BST was developed in response to the litany of concerns over the use of synthetic growth-promoting drugs and antibiotics. However, BST was condemned even before it was released.

I would like to return to the issue of fruits and vegetables and try to answer questions about how society is to deal with maintaining agricultural production if any type of artificial chemical is automatically to be condemned. Readers by now should have made up their minds as to whether pesticide residues at trace levels on produce are a benefit or a threat to their health. I hold the position that there is *no direct evidence* that they are harmful. I do *not* take the position that trace levels of pesticides are inherently beneficial to your health. The evidence presented was that eating fruits and vegetables, even with pesticide residues, is beneficial. Ecological arguments about synthetic chemical use are a completely different issue. Complete elimination of pesticides could be desirable for other reasons if produce availability, affordability, and quality were not to suffer.

For the sake of argument only, let us take the position that we should eliminate pesticide usage in agriculture. As argued for BST, the use of peptide hormones is superior to the widespread use of less specific antibiotics and would be consistent with the beneficial goal of decreasing the selection pressure to develop antibiotic-resistant bacteria. This is good. Similarly, there might be better alternatives to the use of synthetic chemicals to control pests. Agricultural chemicals are expensive to use and may be hazardous to applicators and other workers whose occupation brings them into daily contact with high concentrations of the chemicals in a range that is known to produce adverse effects. Also, there is the potential that agricultural runoff may have negative ecological impacts and could adversely affect our aquifers and, potentially, our long-term water supplies. Thus I would agree that, based on these types of concerns, one could have reason to reduce pesticide usage and try to develop environmentally safer alternatives. For the sake of argument only, I will even assume that chemicals may be producing toxicity to the general public and thus should be eliminated. What then are the alternatives?

Earlier in this book, I mentioned one such movement, IPM (Integrated Pest Management), where the goals are to reduce pesticide use by employing better management techniques and biological principles to reduce chemical usage. This approach is designed to control insect and weed pests by better understanding the biology of agriculture and only using small amounts of specific pesticides or herbicides when required. This refreshing philosophy, used in conjunction with the chemical industry's development of safer pesticides and with efforts made by farmers to control production costs, has optimized chemical pesticide usage and will continue to do so without adversely affecting agricultural production. Note that this rational approach is different from pure organic farming, which does not use any synthetic chemicals.

This book is not about how to raise fruits and vegetables. It is about chemical food safety and the rational application of toxicology to the process. If organic farming would allow for sustainable agricultural production so that *all* people could eat and afford a healthy diet, then so be it. I support this effort, as long as one's motive is not solely driven by an irrational chemophobia or some alternative political agenda. These motives should be debated on their own merit, independent of toxicology. Similarly, IPM is an important development in agriculture that will allow for the optimal use of pesticides and further decrease of chemical residues. This is a good development and should be supported. As will be obvious in the final chapter, I believe that this type of thoughtful approach to the use of pesticides or antibiotics, which is firmly rooted in sound biology, must be adopted if we are to further improve the human condition in the twenty-first century. We need all the tools at our disposal.

There is another approach to reducing our dependence on agricultural chemicals that is consistent with the goals of the above movements and holds the promise of substantially reducing organic chemical usage. This is biotechnology and plant bioengineering. Part of the reason that biotechnology was pursued as a commercial enterprise was that it promised to reduce chemical usage by engineering plants to be inherently resistant to pests, such as insects, fungi, or weeds.

If a person truly believes that chemical usage should be reduced for whatever reason, then that person should be an ardent supporter of modern biotechnology, for it holds the promise of ultimately eliminating all chemical use. As I argued in the previous chapter, if one is fearful of chemical residues and antibiotic resistance, then he should champion BST. However, this is not happening, and the products of biotechnology are attacked as vehemently as pesticides ever were. These should be two separate issues with the only common threads being that they are products of technology and that scientists developed both. If this issue is actually a frontal attack on science and technology, then it should be billed as such. The arguments should be philosophical and social and should not misuse science to induce public hysteria. Biotechnology is a young and struggling industry that has the promise of making significant improvements to public health. It must be judged solely on its merits and failures. Each project should be assessed independently and not condemned outright because of a pervading antitechnology bias.

In this chapter we will address a few exciting developments that are occurring in biotechnology and will attempt to assess what the potential adverse effects could be. There are numerous disciplines that fall under the broad category of biotechnology. We will focus on genetic engineering applied to plant production because it is most closely related to the use of synthetic pesticides. The issues related to drugs developed using biotechnology procedures, such as the recombinant DNA production of BST, were introduced in the previous chapter. These efforts will continue to explode as nonpeptide drugs, such as "antisense" nucleotides which directly regulate genetic functions, appear on the market.

The basic strategy of plant biotechnology is a refinement of plant breeding and hybridization techniques practiced since the dawn of agriculture. The principle is that of natural selection and is the same that occurs in evolution. Plant breeders artificially select individual plants with desirable characteristics rather than letting the environment select the hardiest plants for survival. The desirable characteristics differ between individuals because each of different genes produces different inherited properties (phenotypes) in the new generation of plants.

Plant breeding thus allows selection of hybrids whose phenotypes are best suited to the breeder's needs. These may be qualities such as the ability to grow in direct sunlight or at a higher ambient temperature, a reduced need for specific nutrients, a better root system, production of fruit with thicker skins to facilitate transportation, or some property that confers resistance to an insect pest. When only plants that have inherited these specific phenotypes are selected, the genes controlling these properties are likewise chosen. When the selected plants are grown, most will also share these new characteristics because their genes will now code for these qualities.

Most readers are familiar with this process of plant breeding and hybrid selection. It is this selection process that is responsible for most of the agriculture crops in production today and is a major factor in the increased productivity of modern agriculture. Anyone with a garden has seen the results of selective breeding. So what is the problem?

This process only allows for selecting genes by selecting specific chromosomes (packets of genes) that get inherited in each new generation of plants or animals. A quick biology review is needed to understand this process. The concerned reader should consult an encyclopedia, biology text, or even transcripts from the O.J. Simpson trial for further details. There are four levels of genetic organization we need to understand before we go further. The unit of genetic material to consider is DNA (deoxyribonucleic acid). DNA is composed of two complementary strings of molecules twisted around each other whose building blocks are four different nucleic acid bases: adenine, guanine, cytosine, and thymine (A,C,G,T). Trios of these bases are called codons. Codons are an alphabet for identifying specific amino acids which are the building blocks of proteins, the biochemical components responsible for the structure and function of all cells. Thus a specific protein made up of a defined sequence of amino acids could be considered a code that could be translated into a codon counterpart. This DNA code is analogous to a computer software program which will direct the hardware, the cell in this case, to manufacture new proteins.

In the last chapter, we discussed BST, a peptide hormone composed of some 191 amino acids. Strands of DNA located in bovine cells contain the codon sequence to produce BST. That codon sequence, when combined with other codons used to regulate the DNA copying process, makes up a gene. Genes code for proteins and consist of DNA strings that define a specific codon list. The codons on this list define the specific sequence of amino acids that constitute a specific protein. Genes are then sequentially linked together to form the microscopically visible units of inheritance, chromosomes. Every species has a certain number of chromosomes that, when expressed in cells, code for proteins and thus defines the biology of the species. In humans, all of our chromosomes in a single cell, coding for all of our genes, consist of some

2.9 billion base pairs. This is a very long string. Each chromosome contains millions of genes whose expression makes us the unique individuals we are. The human genome was sequenced in 2001 and published in what will be always considered a landmark issue of *Science*. The genetic code of other species has likewise been determined, and efforts are underway to sequence the DNA from numerous species, both plant and animal.

The order in which the genes are linked may be in groups related to some function or may be more random. For complicated phenotypes that require multiple proteins for coordinated development, genes may be grouped together and controlled as a unit. Some genes required for a specific trait may be located on different chromosomes. Breeding selects for chromosomes and thus when a trait is selected, all the genes come along for the ride. Although this is an oversimplification, biotechnology, in the guise of genetic engineering, essentially inserts specific genes into a chromosome, and these genes code for specific proteins. The new plant, termed transgenic since it now has a gene not originally present in this species, then expresses this inserted gene and produces the protein.

The unique aspect of DNA is that four molecules can actually convey all of the information needed that define life. It does this by the codon alphabet. This codon alphabet is "translated" into the appropriate amino acid alphabet through another type of nucleic acid, called ribonucleic acid or RNA. As a note, viruses essentially do the same thing as biotechnologists. They insert their DNA into the DNA of the host they are infecting and cause the host cell to produce viral-DNA products, which are the proteins needed to make viral components. This ability of viruses to transfect cells is used in transgenic technology where specific genes desired to be inserted into a chromosome are delivered by a bioengineered virus. It is also similar to a mechanism whereby a drug-resistant bacteria can transmit the genes conveying resistance to another bacteria. RNA viruses can take some shortcuts and even be more efficient at invading host cells and usurping control of the genetic machinery. As will be seen in the final chapter, treating viral infections is an extremely complicated task since we must kill organisms that have taken control of genetic machinery in our own cells. They are not like insects or bacteria that have an independent life with unique biochemical targets; instead they are more analogous to a computer program occupying the very computer that we must use to eliminate them.

The study of the structure and function of DNA, as well as its applications to biotechnology, is termed *genomics*. The study of the resulting proteins that are produced is termed *proteomics*. The cracking of the human genetic code has led to great investments into attempting to manipulate human DNA to cure disease by either deleting genes which produce adverse effects or adding genes to produce products that are lacking in the diseased genome, for example, insulin or growth hormone production. The approach of twentieth century science was to administer the missing protein by injection; the promise of twenty-first century science will be to insert the gene that codes for this protein and let the cell do the rest.

DNA is truly analogous to a computer program consisting of instructions to accomplish a task. Computers can be the same, but they might do very different things depending on what the program commands. Biotechnology uses the same process and "slips" a specific gene that codes for different proteins into the plant chromosomes. This was first employed by creating a genetically engineered strain of corn that produced a specific enzyme in the root of the plant. This specific enzyme was deadly to a major pest of corn, the European corn borer. This insect is attributed with causing $1 billion in annual crop loss in the United States alone. The enzyme, specifically named Cry9C, belongs to a class of poisonous proteins known as delta endotoxins. The genetic information to code for this poison is obtained from the bacteria *Bacillus thuringiensis* (Bt for short). Such genetically modified corn is thus termed Bt Corn. As a result of this development, chemical pesticides should not be needed to kill the European corn borer. Additionally, elimination of the borer may decrease opportunistic fungal infection which invade the weakened plant that themselves may be associated with mycotoxin production. Since the Bt toxins are proteins, they, like the rest of the plant, will degrade in the environment after the corn is harvested.

Similar strategies have also been applied to potato, cotton, and soybean crops, conferring resistance to insect pests as well as inserting genes that make them resistant to the effects of herbicides. This allows farmers to spray fields with a chemical such as Roundup® without killing the crop being raised. In 2001, it is estimated that approximately 26% of corn acreage, 68% of soybean, and 69% of cotton raised in the United States will use genetically modified seeds. Some products are being designed which increase the production of nutrients and anticarcinogenic substances naturally found in vegetables in an attempt to increase their benefits even more. Food crops such as sorghum and millet, staples for much of starving Africa, are difficult to grow because of susceptibility to the parasitic Witchweed, *Strigia*. Bioengineered sorghum resistant to *Strigia* would significantly reduce starvation. Arguably the most promising bioengineered plant, one that symbolizes the potential benefits of transgenic technology, is "golden rice." This is rice which is genetically engineered to produce vitamin A and which also confers a golden hue to the resulting rice. Rice is the staple of much of the developing world but lacks vitamin A. It has been estimated that up to 2 million children die each year from this deficiency. Another half a million may go blind.

Are these products safe? Critics suggest that inserted gene products may be toxic to humans or other beneficial insects eating these crops. However, compared to the older practices of breeding, this is unlikely since only specific genes are being inserted. With traditional breeding, whole chromosomes were being selected that may have carried adverse genes for the ride. We discussed some such toxic potatoes and celery in Chapter 3.

The first transgenic plant ever introduced into the American market was the Calgene FlavrSavr® tomato, which underwent voluntary FDA safety testing and approval. This was an extra step taken by the company to gain the public's confidence. The genetic engineering involved in this tomato is the insertion of a gene which results in delayed ripening so that the skin stays hard for transportation but the tomato still tastes fresh. Consumer acceptance of this was problematic. The next major introduction was Bt corn. This is the same corn, produced by Aventis CropScience and marketed as StarLink®, which was detected in corn taco shells in 2000 and led to public hysteria similar to the pesticide scares addressed before. Bt corn has undergone extensive safety testing under EPA guidelines to ensure that the resulting corn does not contain other properties that may be harmful to human consumption. This safety assessment was repeated and led to reapproval at the end of 2001. Issues relating to the safety of such products will be discussed later in this chapter.

A fascinating twist on plant bioengineering is the insertion of genes that code for human peptide drugs such as insulin and growth hormone into plants. Currently, bacteria form the basis of commercial biotechnology, that is, their genetic machinery are the "hardware" that implements the DNA commands. Genes of bacteria, such as *Escherichia coli,* are used to produce our drugs. This is how BST is produced. Genetically altered bacteria form the basis of the current biotechnology industry. They show promise in many areas, such as improving the environment, as evidenced by the development of bacteria capable of breaking down oil and thus safely mitigating the ecological effects of oil spills. Similar approaches are being developed to break down and destroy those persistent chlorinated hydrocarbons, such as PCB, which have polluted our environment. Plants are being bioengineered to be resistant to metals in soil, allowing reclamation of now essentially sterile land. Use of plants instead of bacteria as the genetic machinery for gene expression would allow us to mass produce drugs in crop plants. Thus potato and corn crops could be grown using plants that have insulin genes inserted, allowing techniques of agriculture to be used to harvest the next generation of human drugs. Plants can be made to synthesize novel oils, potentially replacing our dependence on petrochemicals. In fact, some day chemical plants may be located on farms raising tobacco, rapeseed, or soybeans rather than the factories of today. Plants can easily be grown in polluted sites to degrade persistent toxicants. Finally, plants are even being engineered to produce vaccines against a host of human diseases. The potential for plant biotechnology to improve human health is staggering.

These developments, although appearing dramatic and capable of drastically changing agriculture and the pharmaceutical industry, pale in comparison to the long-term possibilities. At this point, only genes coding for simple proteins have been transferred. As we indicated, sometimes more complex traits may involve multiple proteins and thus larger gene complexes. Bioengineers are now working on transporting traits, such as resistance to specific diseases or drought conditions, from plants possessing these traits to susceptible food crops. Rice and potato plants resistant to endogenous viruses may be produced which would allow these crops to be raised by local farmers. Genes for resistance to specific herbicides may be used so that crops are resistant while weeds are not, thereby increasing the specificity and efficiency of herbicide usage. Nutritional value of plants may be improved while the levels of the endogenous carcinogens discussed in earlier chapters are specifically reduced. The nitrogen-fixing properties of legumes may be incorporated in other crops such as corn so that the use of nitrate fertilizer could be drastically reduced. This list could go on forever.

However, there are other benefits. As we gain a further understanding of what genes control specific traits, we will also learn about the cause and potential treatment of numerous diseases. We are already understanding more about the control and regulation of gene expression secondary to exposure to certain toxic chemicals. This research has allowed development of the antisense drugs which directly control gene expression. Such research can only increase our ability to prevent or treat these diseases.

The products being produced today are only the tip of the iceberg. The drugs are inherently safer than the previously utilized organic chemicals because they are more specific in their range of activity and thus produce fewer side effects. Since many are peptide hormones or short pieces of antisense nucleic acids, they have short biological lives and do not accumulate in the body or persist in the environment. If eaten by humans, they just add to the caloric and protein value of the meal. Many of the specific chemical attributes that were cause for concern in the older generations of chemical drugs and pesticides are not shared by these newer products. In fact, their major problems now are expense and difficulty in developing delivery systems. Expense could be drastically reduced if crop plants rather than bacteria were used to produce the drugs. The delivery problem for human drugs is still not solved; however, a number of new strategies have been developed. Of course the ultimate approach to solving these problems may be to transplant the gene's coding for the needed proteins directly into human cells. Similar strategies have been used in cows to allow them to produce hormone drugs in their milk. There are ethical, moral, and religious concerns with these approaches. This is the topic of another book!

Based upon this brief introduction to a small part of the world of biotechnology, I hope that the reader can now appreciate the potential promise it holds for reducing or even completely eliminating the need for synthetic insecticides or herbicides in the future. If chemophobia continues to take hold in the minds of society, then biotechnology should be the logical alternative to develop approaches to replace pesticide use in agriculture. The potential also exists for reducing the concentrations of endogenous plant toxins—that is, all those nasty poisons, like caffeic acid, that we discussed in Chapter 5—and boosting the level of beneficial vitamins and compounds, such as antioxidants, in fortified produce grown for humans. Despite these promises, biotechnology is consistently attacked in the same breath as chemical pesticides. Because of the public's perception that all technology may be tainted, biotechnology already has a stigma in the public's eye. Before I address what I perceive to be the overall reason for this concern, what are the potential problems associated with plant biotechnology? What are the risks, and how much do we know about the potential hazards?

There are two levels of concern relative to transgenic plants. The first is in the realm of toxicology and is similar to that of any pesticide product, which is to insure that the plant itself is nontoxic to humans. The second relates to the genetics and concerns whether the genes that have been transferred to these new hosts could ever leave and transfect other plants or even different life forms. The first concern is in the realm of food safety, and I believe can be addressed with existing technology. The second is in the realm of ecology, and data is lacking to address many of these issues. Again, I must stress that the coverage in this book is focused on chemical food safety and not ecology. There may be reasons that bioengineered plants should not be introduced into the environment. However, arguments about their inherent food safety should not be used out of context.

The toxicology of transgenic plants relates to whether the proteins expressed by the inserted genes—the compounds which kill the plant pests—are dangerous to humans eating them. Most of these are protein products expressed in areas of the plants that are not normally eaten by humans, but the genes are present in the edible components. Because they are proteins, as discussed in the BST chapter, most would be digested along with the rest of the plant. However, some work has indicated that some forms of Bt proteins in corn are more heat stable and resistant to digestion than other proteins. This would allow them to persist in human digestive tracts. Since they are proteins, they still could not be absorbed. However, as will be discussed below, they could prove to be allergens to sensitive individuals.

It is conceivable that new, nonprotein compounds could be produced by these hybrid plants that could also be toxic to humans consuming them. Alternatively, it is possible that some manufacturers could get sloppy and inadvertently carry other genes along which would contaminate the genome. To test for this, existing regulations require that these plants undergo safety testing. These well-established toxicology protocols, which are used to weed out toxic drugs and pesticides, would detect bioengineered plants that expressed toxic proteins and chemicals. Curiously, plants produced using classical breeding techniques, which have the potential to carry along some true deleterious genes, are not required to be tested. Does this treatment promote the public health?

Biotechnology is more selective in the manner in which plant characteristics are controlled. Because new products must undergo batteries of toxicity tests before regulatory approval, the products approaching market are safer than most other plant-derived produce developed using conventional agricultural techniques.

There is a concern shown in recent evidence with Bt corn that insect pests targeted by these gene products may develop resistance to the products, thus making them ineffective. This is identical to the resistance seen against normal organic pesticides. However, this is part of the natural process of adaptation and natural selection and is no different from the adaptation seen over eons of evolution or with pesticide usage. One strategy for handling this is to build different combinations of such toxins into the plants to discourage development of resistance, a strategy that is conceptually similar to that used in preventing antibiotic resistance in bacteria or drug resistance in AIDS.

One reasonable toxicologic concern is whether some of the protein products of transplanted genes may be allergens in certain individuals. This has been observed in many natural plant products. The ruggedness of the Bt protein to destruction by heat increases this concern because it suggests that such allergens might survive food processing, much as the protein allergens found in peanuts. This issue was the fuel that fired the debate over safety of StarLink® corn. Allergies are always a thorny issue and have generally not resulted in preventing products from being marketed unless a large proportion of the population is affected. The concern with Bt protein Cry9C is not that it is a known allergen but rather that it is heat resistant. The molecular structure of Cry9C is not similar to known allergens and thus does not raise a flag. EPA's review of the data did not suggest that this would be a problem because the levels of Bt protein are orders of magnitude less than necessary to sensitize individuals to other compounds. This is not the situation with peanuts, where the allergen is an integral component of the peanut itself. The StarLink® scare was initiated when the DNA coding for Cry9C was detected in taco shells, not the protein itself, which is primarily expressed in roots. Similar to most pesticide scares covered earlier, detection and not exposure is the driving force. One bioengineered plant, soybeans, was pulled from development when a known allergen from the Brazil nut was expressed. Thus the potential for allergen expression in transgenic plants is real but can be identified and predicted based on known science. Toxicologic tests to detect allergens have not been perfected. This issue is no different from that faced for many compounds. Science and medicine are developing rational approaches to effectively deal with these issues, as they are also major stumbling blocks to the development of protein-based drugs. However, most existing foods are not tested for safety nor allergenicity. As discussed in Chapter 5, we know that they contain known carcinogens and other toxic chemicals. We must be cautious to apply our ever-sensitive toxicologic tests in a rational manner and not be hostage to hysteria.

There are some potentially serious scientific issues with bioengineered food that should be addressed. Some workers have expressed concern that transplanted genes in crop plants, which confer resistance to pests or herbicides, may get transferred to wild relatives of these crops and produce so-called "super weeds." In fact, the opposite often happens, and the crop plant itself may reject these foreign genes and lose its new traits upon continuous propagation. There has been some concern that certain genes used to "mark" transplanted genetic material in the process of transferring them to plants—for example, resistance to the antibiotic kanamycin—may become incorporated into the genetic material of soil bacteria, making them resistant to this drug. This is an extremely hypothetical event that has no solid basis in science. Some are concerned that genetically altered plants may spawn novel viruses, another occurrence that is considered too theoretical to even postulate serious hypotheses.

Ecological concerns are real as we do not know what may occur with the widespread propagation of transgenic plants. There were initial fears that Bt corn was toxic not only to the European corn borer but also to the monarch butterfly larvae that may eat milkweed leaves laced with Bt corn pollen. However, EPA studies dismissed this threat. It is possible that other insects, not intended as targets of bioengineered proteins, may also be affected. It is absolutely essential that these ecological issues be addressed. However, these are not food safety issues. The same concerns occur with the widespread use of organic pesticides relative to runoff and water pollution. Both must be addressed on their own merits.

There is another aspect of this entire debate relative to food safety that I have not yet addressed and, in fact, is one where biotechnology has great potential. There is consensus among food safety professionals that 98% of all *real* foodborne illnesses are related to bacterial food poisoning. There are numerous causes of such diseases that have proven very difficult to eliminate. These include meat and produce contamination with bacteria such as *Staphylococcus, Salmonella, Eschericia coli* and *Clostridia botulinum*. These organisms and the diseases they cause have killed more people in the United

States than are predicted by the most extreme estimates to die from cancer secondary to pesticide residue contamination. These are serious problems that have always plagued mankind because it is extremely difficult to detect bacterial contamination during meat processing. However, biotechnology has developed some rapid screening tests to improve the detection of contaminated food before it reaches the consumer. This is another spin-off that has minimal risk and almost immediate benefits.

A major area of concern is more problematic because the issues raised are speculative and are often rooted in fiction rather than fact. Genetic engineering raises all types of bizarre concerns that have no basis in reality. However, the public often cannot detect where reality ends and fiction begins. This is partially a problem of the scientific establishment for not making an increased effort to better educate the public. Instead, the public is left with bizarre impressions of biotechnology gone awry. Who can forget the arrogance of the bioengineers in *Jurassic Park* who thought that they understood dinosaur genetics sufficiently to clone them for that doomed amusement park? When they were pressured to develop a "product" sooner, they spliced frog DNA to make up for missing dinosaur sequences not realizing that they now gave dinosaurs the ability to change sexes and reproduce. *This is science fiction and not reality.*

We have all read popular accounts of plant biotechnology that deliberately scare the public by drawing analogies to what is good science fiction. Just the use of the term "Frankenfoods" to describe transgenic crops congers up the specter of the fictional transgenic monster, Frankenstein. This approach has been so successful in Europe that transgenic plants are prohibited from entering the human food supply. Recall from our earlier discussions, that the genetic control of complex traits requires the expression of a number of proteins, with genes often residing on multiple chromosomes. Transplanting the genes that code for a few proteins will not make our bioengineered corn sprout legs and chase us around the farm. There are limits to biotechnology, and for the foreseeable future, we will be limited to expressing a few new proteins. Similar hysteria was associated with the fear of mutations developing from radiation exposure in the 1950s. Have you seen any monster caterpillars lately? For that matter, have you seen any mutant turtles developing out of toxic chemical waste? Or have evil robots emerged from computers, as predicted from some early "computerphobes?" Science fiction has a place in society, but it is not in the halls of regulatory agencies.

There are dangers of biotechnology and pesticide use in agriculture that relate to other aspects of society such as ecological stability, regulation of human population growth, the control of emerging diseases, and the overall quality of life in future societies. These are separate issues that will be discussed in the final chapter. Finally, some individuals have moral, religious, ethical, and emotional concerns about transplanting animal genes into plants and replacing natural crops—a misnomer since most are true hybrids—with transgenic ones. These are issues of bioethics, which must be addressed on their own merit, and are well beyond the realm of science and food safety. The thrust of this book is not to defend or condemn biotechnology. The reason that I address it is that the attacks being made on biotechnology are similar in nature to those launched against pesticides in fruits and vegetables. These are complex problems that defy simple solutions. A similar chapter could be written about the supposed hazards associated with food irradiation. Discussing this process would only cloud the present topic. I was prompted to consider biotechnology when both BST and bioengineered tomatoes were condemned in the same "news" program that was exposing the ills of pesticide residues. All three were seamlessly linked into a single problem. Having worked with peptide hormones, I knew that there was *no* concern for human health. If they were wrong for pesticides and BST, maybe biotechnology was just being condemned by association. Unfortunately, I believe this was the situation.

The products of biotechnology offer society new tools to combat disease and feed the ever-growing world population. Some argue that biotechnology promises a new Green Revolution and may be the only way to feed the world's population in the coming decades. They do not share many of the problems potentially associated with organic chemicals. If used in IPM programs in conjunction with pesticides, the overall effect of each component could be enhanced. There are potential problems which could develop using these new tools; however, they must be systematically investigated using objective science and not discarded on emotional grounds.

Similar to what happens with many new technological innovations, our imaginations are fueling fears that are beyond reality! Some concerned individuals counter this argument with the revelation that scientists have made mistakes in the past, and that they should not be trusted this time around. My answer to this type of argument is to suggest to those individuals who read the earlier chapters of this text that they should seriously consider the question: Are we really that "bad off"? Do we want to return to an era when our life expectancy was half of what it is today and people routinely died from diseases that have been long forgotten? There are risks to any new technologies, but we must face them and not turn our backs. Otherwise, their benefits may never be felt.

> Not enjoyment, and not sorrow,
>   Is our destined end or way;
> But to act, that each to-morrow
>   Find us farther than to-day.
>
> (Longfellow, "A Psalm of Life")

10

Food Security and the World of Bioterrorism

The tragic events of September 11, 2001, forever changed how the world will assess risk relative to deliberate contamination of the food supply by terrorists or other enemies of a country. What was once a far-fetched scenario must now be considered a serious threat. As I write this chapter, news is breaking on the "anthrax letters" transmitted through the mail. Numerous scenarios are being discussed relative to deliberate contamination of our food and water supply. Since this book has dealt with food safety and accidental chemical contamination, it is instructive and timely to review the nature of this threat and whether our government is prepared to handle it. The reaction to the initial letters discovered also underscore the panic which ensues when unknown agents threaten our safety. Fear drives decisions rather than rational thought, a reaction that unfortunately is totally expected.

The field of bioterrorism, as well as biological and chemical warfare, is a world onto itself. The use of chemicals and infectious agents to kill other humans goes back to ancient times. Table 10.1 lists the infectious agents and chemicals that have been previously employed, or postulated to be suitable, for chemical warfare. Table 10.2 is the current list prepared by the Centers for Disease Control (CDC) of select biological pathogens and toxins that could potentially present a biosecurity threat. The lexicon of bioterrorism may at first appear confusing because terms used commonly in other fields have specific meanings. The popular press often further complicates the picture.

Agent is a term used to indicate a specific form of biological or chemical entity that could be employed in a military environment. Biological agents

TABLE 10.1. Biological and chemical agents relevant to bioterrorism and military warfare.

Infectious Agents:	Anthrax
	Plague
	Brucellosis
	Cholera
	Smallpox
	Viral hemorrhagic fever (Ebola, Marbug)
	Tularemia
	Q fever
Biotoxins:	Botulinum (Botulism)
	Ricin
	Staphylococcal enterotoxin B
	Mycotoxins (T-2, Nivalenol, Alflatoxin)
Chemical Agents:	Vesicants: Sulfur Mustard (HD), Lewisite (L)
	Nerve Agents: Sarin (GB), Soman (GD, Tabun (GA), VX
	Phosgene, Hydrogen Cyanide, Arsine, Chlorine

TABLE 10.2. Current CDC list of select biological pathogens and toxins.

Viruses:	Ebola	Crimean-Congo haemorrhagic fever
	Smallpox	Tick-borne encephalitis
	Marburg	South American haemorrhagic fever
	Eastern equine encephalitis	Venezuelan equine encephalitis
	Rift valley fever	Hanta virus
	Lassa fever	Yellow fever
	Equine morbillivirus	
Bacteria:	Anthrax	Brucella abortus
	Clostridium botulinum	Francisella tularensis
	Burkholderia	Yersinia pestis
Rickettssiae–Fungi:	Coxiella burnetti	Rickettsia ricketsii
	Rickettsia prowazekii	Coccidioides immitus
Toxins:	Abrin	Ricin
	Aflatoxins	Saxitoxin
	Botulinum	Shigatoxin
	Clostridium perfringens	Staphylococcal enterotoxins
	Epsilon	Tetrodotoxin
	Conotoxins	T-2 toxin
	Diacetoxyscipenol	

include infectious diseases such as bacteria and viruses, as well as the toxins produced by living agents such as botulinum produced in cases of botulism. Chemical agents are chemicals that are used in a military environment. In other scenarios they may be pesticides. A commonly used "simulant" to study nerve agents is diispropylfluorophosphonate, an organophosphate insecticide too toxic for use as a pesticide but not toxic enough to be used as a chemical warfare agent. In fact, the development of nerve agents in the early twentieth century occurred in laboratories whose initial focus was the development of pesticides. The study of protection against biological or chemical attacks is termed "biosecurity" or "chemical defense," depending on the agents discussed. The protection of food against both biological and chemical attacks is termed "food security."

The history of biological and chemical warfare is lengthy and texts on the subject should be consulted for further details. The focus of this chapter is an extension of the book, foodborne diseases. However, a brief review of this field is instructive. Most of what is presented here relative to terrorist attacks in food, and which has been written about by other authors, has never occurred. This must be viewed as pure speculation extrapolated from the use of chemicals and infectious disease in a military environment. Such speculation, once considered pure fantasy, now has a chilling reality to it based on recent terrorist attacks throughout the world.

In ancient and medieval times, diseased animals would be thrown into cities under siege to spread infectious disease amongst the enemy. Corpses who died of the bubonic plague were catapulted over the ramparts in battles to kill the trapped inhabitants of the town under attack. Leonardo da Vinci reported the use of saliva from rabid dogs as a biological warfare agent. In fact, the use of smoke to disperse the enemy or provide cover for troop movements, throwing boiling oil over the castle wall, as well as the use of tear gas in riot control, are all technically forms of chemical warfare agents. Another form of military use of chemicals, one that the United States was accused of using as chemical warfare in Vietnam, was the herbicide defoliant Agent Orange. This was sprayed in battlefield environments to destroy vegetation where the enemy was known to hide. These types of chemical attacks are not generally considered true "chemical warfare" as they do not inflict the extreme casualties, and battlefield chaos and hysteria, that the prototype gases and nerve agents do.

The first use of chemical warfare on a large scale was in World War I during which some 1.3 million casualties—exposures resulting in clinical symptoms, not deaths—were reported. Seven percent of these were fatal. However, this must be viewed relative to the total casualties of 28 million resulting from all other forms of warfare. Additionally, of all deaths (by all causes) for all forces in World War I, only 1.3% could be attributed to chemical agents. The first gas attack occurred on March 9, 1918, when the German army fired mustard gas shells against Allied forces. Agents used included the vesicants (blister-forming) sulfur mustard and arsenic-containing lewisite, as well as the gaseous chemicals phosgene and chlorine. Vesicants are deadly both from inhalation exposure and dermal contact. Breathing sulfur mustard is extremely irritating and results in a traumatic death. On the other hand, phosgene provides a delayed reaction and is not immediately detected after chemical exposure. However, a few hours later, severe pulmonary edema occurs, which is often fatal. Phosgene (carbonyl chloride) as well as chlorine are commonly used industrial chemicals that are widely available as raw materials in chemical manufacturing.

Once the use of chemical gases became widespread in World War I and adequate protective measures were taken (e.g., gas masks), casualties dramatically decreased. However, efficiency of soldiers is drastically reduced. Since this occurs for both sides of a conflict, some military experts do not believe that the use of chemical agents provides a distinct strategic advantage. What they do provide is a fear factor that is not shared by use of common munitions. The highly potent nerve agents were never used in World War I since they were developed after the Armistice in the thirties.

Chemical warfare agents were not used in World War II, although both sides had the capability. Reports have surfaced that chemical agents were used in experiments conducted on concentration camp prisoners, by both the Third Reich and Japan. An accidental exposure to mustard did occur during World War II when a tanker was sunk and mustard leaked and contaminated the waters of an Italian port. Survivors rescued from the sinking ship developed classic skin lesions associated with sulfur mustard.

Since World War II there have been sporadic incidents where chemical weapons were used. The most famous was Iraq's use of sulfur mustard, and possibly the nerve agent tabun, in the Iran/Iraq war during the 1980s. One of the worst reported incidents was Iraq's use of chemicals against its own Kurdish civilians in Halabja, where 5000 people were purported to die. This was the worst use of chemical agents since World War I. Accusations of chemical agent use were made in Southeast Asia (yellow rain) and against the Soviet Union in Afghanistan (black body agents). This must be differentiated from those who accuse the United States of using chemical agents in Vietnam, since these accusations concern the use of the herbicide Agent Orange discussed above. Note that although Agent Orange exposure has been discussed previously in terms of the accidental contamination with dioxin (a contaminant produced in herbicide manufacturing), this herbicide was intentionally used to kill only foliage, not people.

Large stocks of sarin, cyclosarin, and sulfur mustard were found by United Nations inspectors in Iraq after the Gulf War. Similarly, nerve agents were detected when coalition forces destroyed the munitions dump at the Kamisiyah depot in Iraq at the end of the Gulf War. This incident has received great attention since some purport that exposure to low levels of sarin released in the explosions may have contributed to the Gulf War illness among coalition forces. This issue is still being debated.

The use of chemical agents by terrorists has a more recent and much more abbreviated history. It has been suggested that Osama bin Laden and his al-Qaeda training camps in Afghanistan specifically instructed terrorists on the use of chemical and biological agents. Currently, the developing saga of anthrax letters is a case where large number of casualties may not occur; however, individuals may be targeted, business disrupted, and the rest of us become unnerved. The most documented case of a terrorist organization's use of a chemical warfare agent involves the group, the Aum Shinrikyo, a Japanese religious cult who used a nerve agent in a Tokyo subway, resulting in 12 deaths. This group exposed many civilians to anthrax, although their attempt at using this agent was not successful. This latter failure brings to the forefront the concept of militarization of biological and chemical agents, which is the development of effective delivery systems to lethally affect large numbers of individuals. Luckily, this is not an easy task and usually involves organizations normally associated with nation-states. However, many commercially utilized chemicals such as phosgene, chlorine, or arsine would be effective weapons if properly dispersed in their commercial form.

The focus of our discussion is on the use of such agents in the food supply. Unfortunately, there was actually a deliberate contamination of food with a biological agent that afflicted at least 751 people with foodborne illness. A religious cult, the Bhagwan Shree Rajneesh, infected salad bars in an Oregon town in 1984 with salmonella, hoping to reduce voter turnout in a local election so a candidate favorable to the cult would be elected. The cause of this food-poisoning incident went unknown for a year until an individual alerted authorities to the event and public health officials and law enforcement agencies determined the cause. Thirteen individuals were intentionally poisoned in another incident in 1996. To date, these are the only known incidents on bioterrorist attacks using food as the delivery system. Numerous hoaxes and threats have been made or reported, but no additional confirmed attacks have ever been validated.

Biological agents which could be adaptable to food contamination include cholera (caused by *Vibrio cholerae*), Q fever (*Coxiella burnetii*), botulinum toxin, aflatoxin, ricin, and Staphylococcal enterotoxin B. Additional agents include those that have been associated with food poisoning by accidental means, including *Salmonella* and *E Coli*. The use of these agents is a function of how many individuals a terrorist would want to target. The smaller the target, the easier the task and the less sophisticated the delivery system needs to be. The Oregon case cited above clearly indicates this. Poisoning a single restaurant or cafeteria is one thing; distributing a toxin widely throughout the food supply is quite another. One of the major factors that detract from the efficacy of a point source contamination of the food early in production is that of subsequent dilution and processing farther along in the food production process. For example, even if one dairy's milk supply were contaminated with a chemical agent or biotoxin, subsequent dilution in the tanker truck by the milk from other dairies would reduce the toxic concentration. This same logic explains why milk from a single cow contaminated with drug or pesticide residues essentially has no public health effect. However, its presence alerts regulators to the conditions operative on that specific farm that would allow this milk to be produced. Unfortunately, the opposite is true for an infectious agent; however, the pasteurization process in milk minimizes this threat.

Biological agents have not been difficult to acquire in the past and anyone with a rudimentary knowledge of microbiology could probably cultivate sufficient infectious doses to cause serious harm. With the events of September 11 bringing such scenarios to the forefront, it is very likely that sources of such agents, such as biological supply houses, will come under increased scrutiny and control as many facets of our open and naive society are altered. In fact, recently passed federal legislation now requires that agents listed in Table 10.2 be registered with the proper authorities.

Frightening scenarios associated with the use of biological agents are imagined when the advances of biotechnology and genetic engineering, discussed in the last chapter, are applied to this obscene science. It must be stressed that there is no evidence that this has occurred in the context of terrorist attacks on any country. Such modifications could include conferring antibiotic resistance to an agent such as anthrax or increasing survivability of such agents when exposed to normal disinfectants or altered environmental conditions, such as temperature or arid conditions. There has been speculation that nation states, such as the former Soviet Union, experimented with such technologies. Some of these laboratories are located in now independent states bordering on Afghanistan, and their security has been known to be "less than adequate" immediately after the collapse of the Soviet empire. Similarly, speculation has surfaced about Iraq and Libya sponsoring such biological warfare experiments. Construction of such doomsday biological agents are probably more the realm of fiction because the current incidences of terrorism that have afflicted this country since the September 11 attacks on New York and Washington are characterized by low technology and relatively minimal expense for the attackers.

It should be brought to the reader's attention that another approach is not to directly contaminate the food with the intent of killing human consumers but rather to kill food-producing animals in an attempt to destroy the infra-structure of our food production system. Deliberate introduction into this country of Foot and Mouth Disease (FMD) would have grave economic con-sequences and disrupt normal production and distribution of food as it did in Great Britain in 2000. Other animal diseases would have similar effects even though they do not produce disease in humans. These are really incidences of economic warfare.

Our focus in this book must remain chemical food safety, and thus what are the chemical agents that could be employed in a terrorist attack on our food supply? The biotoxin most suitable for dissemination in food is botu-linum toxin, a natural agent of food poisoning produced by the anaerobic bacteria, *Clostridium botulinum*. Anyone who has even attempted to bottle or can their own tomatoes is familiar with this poisoning. Botulinum toxin, of which seven antigenic types designated A–G exist, attacks the nervous system and may produce a fatal paralysis after ingestion of less than a millionth of a gram. Curiously, this biotoxin is one and the same as the "Botox" that is used by dermatologists to combat wrinkles and by ophthamologists for treating strabismus or bleopharospasms. The ability to easily culture and produce bot-ulinum toxin makes its large-scale production by a determined terrorist organization a great threat to the public. The Paris police raided a house in 1980 occupied by the terrorist group Red Army Faction and discovered botu-linum toxin being cultivated in a bathtub. The production of such food poi-sons is not a sophisticated process. The plus side to this specific agent is that rapid diagnostic tests are available to detect its presence—if suspected—and heat treatment commonly involved in food processing will inactivate the toxin.

The use of the classic chemical warfare agents in food has not been seri-ously addressed. Many of the agents used so effectively in military conflict—phosgene and chlorine—are gases and are not amenable to delivery in food. The same can be said for the potent nerve agents sarin, soman, and tabun. Their volatility effectively prevents their easy use in food. Depending on the chemical form, sulfur mustard is also a volatile agent and, additionally, rapid-ly degrades in the presence of water. However, this problem is also shared by many commercially used pesticides that have been specially formulated to overcome this limitation. Such techniques as encapsulation, sustained deliv-ery, and the use of additives to promote stability theoretically could be applied to chemical warfare agents to make them effectively delivered in food. There are enough other persistent agents available that would make such an effort unlikely. Secondly, many of these chemical warfare agents are most lethal after exposure by inhalation, much as respiratory anthrax is much more deadly than cutaneous anthrax. Inhalation exposure is not a scenario consis-tent with exposure via food.

Of course, use of any noxious chemical could be considered as potential agents that, if not detected in the normal course of food inspection, could easily be distributed widely throughout the food supply. Consider the repercussions of accidental feed contamination of the fire retardant-PCB in Michigan dairy cattle during the late 1970s or the most recent incident of dioxin contamination of feedstuff in Belgium. These incidents did not result in significant toxicity to human food consumers, but they did disrupt the production and distribution of food much as Foot-and-Mouth Disease did in the United Kingdom. Many industrial chemicals, a list of which is too long to tabulate, could serve as effective agents in this scenario of "agricultural terrorism." Another unknown factor is whether these agents are actually available to be absorbed when present in food. Agents such as aflatoxin and botulinum are clearly capable of being delivered in food. As discussed in earlier chapters, food is a complex matrix possessing many substances which might be capable of either binding to, inactivating, degrading, or competing with the action of a chemical agent administered in food. Such interactions are known widely to occur when drugs are administered in food or simultaneously with meals.

The persistent chemical agents which could be used in a foodborne attack, aimed at poisoning human consumers, include the nerve agent VX, any of the mycotoxins—aflatoxin, T-2 toxin, nivalenol (see Appendix A for discussion of their toxicology)—and compounds such as ricin produced from the castor bean plant. Chapter 5 should be consulted for a full discussion on natural toxins. These chemicals, combined with the biotoxins listed above, could potentially be used in the food supply. This would not be an easy task since many of these agents are toxic to animals and thus might kill the host, preventing dissemination into the food distribution system.

An interesting case that illustrates this point was a natural outbreak of botulism which occurred in California in 1998 and 1999. The Food Animal Residue Avoidance Databank (FARAD) program, which the author is involved with, helped to mitigate this incident. Over 500 cows died after being exposed to botulinum toxin. In one case, the source of the biotoxin was a dead cat that was on top of a stack of hay bails. The cat underwent anaerobic fermentation and produced significant quantities of botulinum. The hay was then mixed into the cattle feed which then killed many animals. The large dose and acute lethality of this exposure prevented milk from the affected cows from being distributed into the dairy system. In the second case, the exposure was more prolonged and probably originated from exposure to rain-damaged hay, a similar condition that may produce aflatoxin in dampened corn. Since cows were dying, veterinarians were able to test milk to ensure biotoxin did not enter the human food chain. However, cows that managed to survive the toxic insult themselves could still have secreted toxin into their milk. Processed dairy products such as butter would be safe, even if toxin were present, because the processing itself inactivates the heat-sensitive biotoxin. One could imagine the scenario where animals were "titrated" with a sublethal dose of toxin and the cows were allowed to continue producing milk. Alternatively, milk in tanker trucks could be directly contaminated with biotoxin as could any fully processed liquid food product. The issue of acute toxicity to animals is obviously mute if the chemical agent is instead used to expose crops of vegetables, fruit, or grain. In these cases, chemical agents could enter the food chain much as we previously discussed with pesticide exposure to crops.

There has been some work done by the military to estimate the toxic concentrations of chemical warfare agents which would be harmful if consumed by humans. This work was conducted when military stockpiles of chemical agents were being destroyed and concerns existed about accidental exposure to neighboring farms. This analysis suggested that the most likely agent with potential for serious exposure was the persistent nerve agent VX since this oily substance would persist on vegetation and could enter the food supply. VX is the most potent nerve agent ever manufactured and, unlike other nerve agents, is not volatile and thus is persistent. The estimates of toxicity were developed by comparing the structure of the nerve agent with organophosphate pesticides, which are of the same chemical class although much less potent. VX is some 1,000 to 10,000 times more toxic than a toxic organophosphate insecticide. Using this logic, safe concentrations in food would be 1,000 to 10,000 times less than allowable pesticide residue levels (see discussion of chemical tolerances in Chapter 6). It is not known how easy it is to extrapolate such values since one attribute that distinguishes organophosphate chemical warfare agents from pesticides is the latter's ability to irreversibly inactivate acetylcholinesterase enzyme activity in affected subjects.

There is no documentation that any serious attempt has ever been made to contaminate a food supply with chemical agents or biotoxins on a scale that would result in massive casualties. The detection of such an attack would be highly dependent upon the type of food selected for the delivery vector (plant or animal products). Many of the common chemical screening tests applied to food products (vegetables, milk, etc.) would actually detect some types of chemical agents such as nerve poisons since they are chemically related to the pesticides that are currently being screened for. As discussed in earlier chapters, the problem with these systems is that only random samples are taken, making the probability low that food actually contaminated with an agent would be detected. The heightened awareness that now exists in the public health community might improve the odds of detection by increasing the sampling and the use of screening assays that could detect multiple chemical types. This ability is well within the reach of modern analytical chemistry techniques. It is ironic that this increased ability to detect chemicals created the chemophobia and hysteria extensively discussed in earlier chapters. This twist actually gives this author great comfort that detection systems can be developed to counter the chemical terrorist threat.

Agents directed against live animals that produce edible products such as milk or eggs would be expected to cause some kind of disease in the animals that could be detected by veterinarians or food inspectors. The increase in awareness of such potential attacks is expected to dramatically change and improve this inspection process as the reality of such a threat takes hold. The use of the HACCP (Hazard Analysis and Critical Control Points) food inspection process, which was designed to test animal products for bacterial contamination by drug-resistant bacteria, could be expanded to test for chemical agents using the sophisticated detection systems that are already deployed by the military. This would be applicable to most food-production systems. In fact, the surveillance systems and extensive focus of research is already in place to combat bacterial food contamination and can be adapted to face this new threat.

The world is now truly a different place. Unfortunately, it is the chemophobic tendencies of the American public that terrorists take advantage of when they launch largely ineffective biological attacks on the public and, through fear, completely disrupt the normal functions of society. Such attacks against our food unfortunately do present a real chemical threat if the worst-case chemical agents are employed. We must adapt our food inspection systems to address this threat and use available tools of analytical chemistry to screen food to protect the consuming public.

The Future

I wrote this book because I am convinced, as are many of my colleagues, that there is minimal risk to human health arising from low-level pesticide residues in our fruits and vegetables. Furthermore, drinking milk from BST-treated cows or eating genetically engineered corn is equivalent to consuming the natural products. *We should stop worrying and get on with life!* In fact, eating fruits and vegetables is healthy. We should be actively promoting and encouraging produce consumption so that we are better prepared to actually reach our maximum life expectancy.

However, no matter how rosy this picture is relative to chemical food safety and bioengineered foods, there are food-related and public health threats that are confronting society, and some people believe could actually threaten the existence of the human species. Some of these are real and need immediate attention. Resources should be focused on them rather than being wasted on many of the nonexistent pseudoproblems outlined in this book. But be forewarned, there is also hype and exaggeration embedded in some of these alarms. We have problems that need to be addressed. Some of the true toxicological issues that face us include how to assess the risk of chemical mixtures or to define the hazards associated with yet-to-be-developed products of biotechnology. In the context of this book, if we are to economically feed the world's rapidly growing population, a real problem that must be addressed by improving agricultural yield, we must continue to develop new and safer pesticides.

There are food safety issues present today that deserve attention. These include bacterial resistance to antimicrobial drugs that result in food contamination by resistant bacteria and human consumption of beef from cows that are carrying infectious prion particles associated with "Mad Cow Disease." These are very serious issues that were not addressed in this book as they are not directly related to chemical food safety and toxicology. They fall in the realm of microbiology and infectious disease. As mentioned earlier, bacterial food contamination is responsible for significant numbers of food poisoning incidents every day. Often these have no long-term effects beyond the acute sickness. In some cases, such as in the very young or very old, serious disease may result when the offending bacteria are resistant to antibiotic therapy. The reason for bacterial resistance is multifold, and blame must be based on over-use of drugs by physicians and animal producers as well as an insatiable desire by patients to be treated with a drug when illness is upon them. As far as bacterial food contamination is concerned, the food production system is developing and utilizing sensitive diagnostic tests to detect resistant organisms before they reach the consumer. However, a large responsibility will always rest with the consumer, who must practice sound hygiene and food preparation practices, which include washing produce and cooking meet adequately.

A serious food safety issue is the discovery of a new class of infectious agents, called prions, responsible for "Mad Cow Disease," or Bovine Spongiform Encephalitis (BSE), in the United Kingdom. Since 1996 there has been increasing evidence linking consumption of beef from cows with BSE to a rare debilitating and fatal neurological disease in humans called "New Variant Creutzfeldt-Jakob disease." Prions are protein particles that can invade nervous tissues after oral consumption. Research is defining how such infection occurs and have implicated initial invasion into a special intestinal cell. Prions cannot live outside of the host and have no genetic machinery to self-replicate. They are literally at the edge of living and nonliving entities.

Prions are not an infectious disease in the classical sense, as they would only be transmitted by eating meat or tissues from an infected animal. Cows acquire BSE from other cows when their feed is supplemented with meat-bone meal from infected animals. Incubation takes years and once the disease takes hold, it is fatal. Other prion diseases are known to exist, including scrapie in sheep, chronic-wasting disease in deer and elk, and transmissible mink encephalopathy in mink. Another human prion disease, Kuru, was transmitted decades ago in New Guinea when human flesh was consumed. All countries have now outlawed feeding meat-bone meal as a nutritional supplement to prevent further outbreaks of BSE or other similar diseases. In the United Kingdom, 4.5 million cows were slaughtered to irradicate the disease. Export bans on cows from infected countries are in place, and excellent progress using the latest techniques of biotechnology has been made in developing and fielding modern diagnostic tests to detect such proteins in animals or their feed. In Germany, 100% of animals are now tested. BSE produced a major public health scare when cows were being slaughtered in the United Kingdom.

In this case, there truly were unknown risks. These are the problems which deserve our attention.

This is a good point to also bring up another food-animal disease that caused havoc in British and European markets, the outbreak of Foot-and-Mouth Disease (FMD) in 2000. It should be noted that FMD is not transmitted to humans and produces no food safety issues. However, it is highly infectious and debilitating to hoofed animals such as cows and pigs, resulting in economic losses. The outbreaks of BSE and FMD in Europe, coupled with recent outbreaks of drug-resistant bacterial food poisoning, have made some individuals truly paranoid about food safety. There is little doubt that this heightened awareness carries over into the field of chemical food safety. However, some issues such as BSE and antimicrobial resistance deserve our attention; others do not.

There appears to be a profound misunderstanding of the role of modern science and medicine in society. It may cause excessive concerns over minor issues while simultaneously providing false security due to the belief that rapid technological progress implies science understands all natural phenomena and can solve all problems. In fact, science may have created many of these problems for itself. Science has an increasing ability to quantify the minute and uncover previously unknown phenomenon; yet it often does not fully comprehend the scientific basis of these newly discovered phenomena.

> Contrary to current folklore, it is not the progress of basic science that chiefly is to blame for many of the large scale functional difficulties we face today, it is rather the absence of some specific fundamental scientific knowledge.
>
> (G. Holton, *Science and Anti-Science*)

The crux of my argument is that throughout history, disease and death have always been associated with life. The development of chemicals and modern drugs in the middle of the twentieth century finally gave humans some effective tools to combat the scourges of diseases that decimated human populations for centuries and caused unimaginable grief and tragedy. Food shortages, which once resulted in horrible famines across the globe, have now largely been eliminated. When they are present today, they are usually due to political and cultural causes, not agricultural ones. The average life span of humans throughout the globe, and especially in Western societies, has dramatically increased. This has created some significant social and political problems that must be handled, independent of attacking the underlying science and medicine that indirectly may have facilitated this growth, and are critical for future development.

The human race remains in a primordial war for survival against all other life forms on this planet. Twenty-first century humans are now engaged in a new battle. The soldiers on both sides of this new war, although on the surface appearing similar to those that fought in numerous battles since the beginning of life on earth, are fundamentally different. The humans fighting are the survivors, the offspring of parents that finally lived through the ravages of bacterial diseases that once killed their ancestors. Medical science, primarily in the areas of pharmacology and immunology, developed weapons, such as drugs and vaccinations, which allowed the fragile human body to survive and thrive in the new habitats. However, the infectious diseases we face off against today are also survivors of our technology and demand that new approaches to battle be developed.

We must realize that modern man never evolved to be capable of populating the entire earth without the aid of a technology fostered by the development of our higher intellect. We are creatures of the African savannah, and most of our biology has evolved over millions of years to survive in this environment. Equipped with these ancient tools, we now face challenges that have only arisen in the last 100 years. The threats existing in this long-gone primordial world were very different from those facing us today. In fact, many of the so-called adaptations that may have aided in the survival of our Stone Age ancestors are detrimental to us in today's modern world. Medicine has developed tools that allow us to cope with these problems. Since most of these modern ailments only kill people in the later decades of life, these problems were irrelevant to prehistoric populations because the majority never lived beyond what we now consider to be middle-aged. As a result, our troops engaged in this primordial battle today are aged compared to those who fought the battle of survival even a century ago.

Humans won the twentieth century skirmishes against our insect and bacterial competitors and created a sustainable agriculture that at the end of the century was beginning to become ecologically friendly. This victory was accomplished through the use of synthetically derived chemical pesticides, disinfectants, and antimicrobial drugs. Our enemies were insects such as mosquitoes that killed millions of people by carrying malaria and yellow fever. Bacterial and viral epidemics, including the plague, tuberculosis, influenza, and polio, decimated entire families and villages. In the case of diseases like leprosy, polio, and tuberculosis, life was horrible for those who were unlucky enough to survive. Lack of sanitation and clean, potable drinking water allowed global pandemics of devastating diseases such as cholera to kill millions of people. Disinfectants and the development of chlorinated water supplies eliminated this problem. Our agriculture at the dawn of the twentieth century required increasing acres of farmland to raise productivity because insect and plant pests prohibited intensive farming practices that would increase the yield of high quality food product from smaller farms. Because of food shortages and lack of storage technology, food poisoning and fungal contamination killed people at will. Herbicides, insecticides, fungicides, fertilizers, and the breeding of hybrid plants changed all of this.

This new battle for survival being fought today is facing fortified enemies. These opponents are the survivors of our chemical and immunological attacks. One must remember that our insect and microbial competitors are fighting the same battle for survival and have new strategies for attack. Where we have intellect that produces modern drugs and vaccinations, they have the ability to mutate, propagate, and reproduce in generation times measured in minutes rather than decades. In the past century our enemies have evolved and undergone natural selection while humans have remained essentially the same as our forebears, except for living longer and developing new technology. The bacteria and insects continue to evolve and develop variants that can survive in the modern environment shaped by pesticides and antibiotics. Humans have forgotten that these battles are ongoing and, through an arrogance founded in an overwhelming faith in modern medicine, they have let their guard down. The new attackers have invaded.

Who are these new adversaries? They are led by viruses such as the human immunodeficiency virus (HIV) which causes AIDS, the hanta viruses, herpes, and the emerging viruses such as Ebola and Lassa fever. They are bacteria, such as *Salmonella*, *Staphylococcus* and *Mycobacterium tuberculosis*, that once were easily killed by a host of antibiotics and are now resistant to most drugs we develop. They are the insect-borne diseases such as Lyme's, Dengue, and West Nile. They are the prions, a lifeform less complex than viruses that can only survive in a host and cause debilitating, and presently noncurable, madness in animals and humans. They are the insects themselves that are resistant to many organic pesticides and which continue to cause destruction of plant life. Why are these threats emerging now?

Part of the problem is the never-ending growth and intrusiveness of the human population across the globe. Throughout history, events such as population migrations, overcrowding, ecological damage wrought by wars or encroaching civilization, and famine have precipitated devastating plagues. Today we are facing the challenges of fighting emerging new viruses that were once adapted to primate hosts and did not infect humans because of geographical isolation. However, the intrusion of humans and the displacement of the adapted primate colonies from primordial rain forests throughout the tropical worlds have created new evolutionary opportunities for numerous viruses, and these viruses are extremely virulent in their new primate hosts. Sexual promiscuity, coupled with the ability of humans to traverse the globe, make what once might have been self-limiting regional epidemics into global pandemics. Evolutionary theory predicts this, and the phenomenon is not new. Forty million people died at the beginning of the twentieth century in the great influenza pandemic. The difference is that the new diseases affecting us at the beginning of the twenty-first century, such as the fever caused by the Ebola virus, are more virulent, or as in the case of HIV, have long latency periods and extreme evolutionary adaptability that allow them to usurp control of their human host's immune systems to ensure their own survival.

Modern man has lost sight of this evolutionary struggle. There is obvious hysteria associated with what appears to be a losing war against these horrible diseases. Richard Preston in his book, *The Hot Zone,* crystallizes the primal nature of this human-viral war, as well as the intrusion of mankind into the balanced ecology of the earth, when he writes concerning the fatal viruses emerging from the tropical rain forests:

> In a sense, the earth is mounting an immune response against the human species. It is beginning to react to the human parasite, the flooding infection of people, the dead spots of concrete all over the planet, the cancerous rot-outs in Europe, Japan and the United States, thick with replicating primates, the colonies enlarging and spreading and threatening to shock the biosphere with mass extinctions. Perhaps the biosphere does not "like" the idea of five billion humans. Or it could also be said that the extreme amplification of the human race, which has occurred only in the past hundred years or so, has suddenly produced a very large quantity of meat, which is sitting everywhere in the biosphere and may not be able to defend itself against a life form wanting to consume it. Nature has interesting ways of balancing itself. The rain forest has its own defenses. The earth's immune system, so to speak, has recognized the presence of the human species and is starting to kick in. The earth is attempting to rid itself of an infection by the human parasite. Perhaps AIDS is the first step in a natural process of clearance.

Anthropomorphisms notwithstanding, I quote this passage for two reasons. The first is to put the topic of this book in the proper perspective. *The threat of dying from pesticide residues in food is insignificant when we face such serious public health threats.* Pesticides, organic chemicals, and biotechnology are essential tools to combat these threats. Secondly, to fight these new diseases, we need new weapons and technology, as well as learning how to use the old ones more effectively. Only modern science will allow us to do this. It is tragic that medicine has not found the cure for these diseases. However, should anyone have expected that this could be accomplished in a mere decade? Molecular biology, coupled with the tools of biotechnology and genomics, are giving us a fighting chance to conquer these new and deadly foes. Our increased knowledge of the mechanisms of toxicity ensures that these drugs and therapeutic approaches will be less harmful than those of only a few decades ago. We have learned the lessons associated with our wars of antibiotics against bacteria, as well as pesticides against arthropods, and realize that we must use more sophisticated strategies, coupled with prevention and containment, to win this war. In some cases in America, these public health techniques, which have worked for decades in controlling other infectious diseases, fly in the face of individual rights and even human compassion. However, we must protect the general public health and some would say, the survival of our species. The actions to be taken are often not easy.

In some ways, these new enemies are similar to our familiar foe, cancer, since we are fighting our own cells, which have either lost control of their growth through malignant transformation or are being occupied and controlled from within by invading viruses and prions. The first complete genetic sequence of a free-living bacteria was published in 1995, while the human genome was sequenced and published in 2001. Ironically, it was *Hemophilus influenze* that was sequenced, the same agent that ravaged the world almost a century ago. This is a benchmark event in the maturation of molecular biology and biotechnology as disciplines. We are making progress! We must realize this and continue to do so.

There are additional diseases that have allied themselves with these infectious enemies in the global alliance against the human species. Some are direct results of our abnormal diets—low in fruits and vegetables but high in salt, sugar, and fat—that lead to heart disease, diabetes, and other ailments of affluence. Many of these diseases have always been present but lay dormant because we never lived long enough to experience them. Mechanical engineers would consider them analogous to innate structural defects that are harmless as long as the manufactured product is used before its expiration date. Modern man has passed this evolutionary age of expiration, and the enemies are the autoimmune diseases and the diseases associated with senescence—generalized organ failure, immunoincompetence, and degenerative ailments such as arthritis and Alzheimer's disease. They are the multifaceted expressions of the new actor playing the role of Longfellow's Reaper, our ancient nemesis, cancer. Today we realize that there is a genetic basis of cancer, and that oncogenic (cancer-causing) viruses are responsible for more cancer than any implicated chemical. Sunlight, not chemical pollution, is the primary environmental carcinogen.

As argued throughout this book, these diseases are not occurring today because of a polluted environment or exposure to trace levels of pesticides. They are occurring because medicine has been triumphant in helping us live longer, and we now are facing the results of this success. If we give in and let our defenses down, the old enemies will just move in and finish the battle.

We are our own worst enemy because we fail to realize that no one person or institution is to blame for the emergence of these modern diseases. They are just the old actors in new costumes playing out their role in another of the never-ending acts of the same play. Because of this profound ignorance, we honestly debate issues such as whether we should remove chlorine from our drinking water to protect that one person in a million, or even one person in ten million, from a theoretical risk of contracting cancer from an organic by-product of chlorination. We will replace this with an almost certain increase of thousands of deaths which will occur from bacterial diseases, such as cholera, which would now flourish in our unprotected water supplies.

This cruel experiment was unintentionally conducted in Peru in the 1980s. The result was a cholera epidemic affecting 300,000 people and causing 3,500 deaths. In the United States in 1994, problems in the water chlorination system of Milwaukee resulted in over a third of a million people being sickened by the protozoan, *Cryptosporidium*. Similarly, as discussed earlier in this book, the removal of DDT from Africa resulted in millions of deaths secondary to a resurgence of once-controlled malaria. The human trials have been done and, in retrospect, the benefit-to-risk ratio favors DDT and chlorine. Chlorine has not yet been banned. Why is there even any debate?

I believe that it rests in the perspective of modern man who believes that technology is actually in control. When disease strikes, it is the fault of science and the technology it has spawned. This logic feels somewhat comforting since it gives us the false belief that we are in control. Our legal system is also poised to find blame even when there is no scientific evidence of a problem.

Just when we are finally developing an effective defense, we are letting our guard down and leaving ourselves open to attack by our ancient enemies. When the problems facing us appear insurmountable, we paradoxically seem to then ignore our best technological solutions and instead retreat to some other system of belief so that we do not have to face reality. Yet reality persists as it always has.

> And I looked, and behold, an ashen horse; and he who sat on it had the name "Death," and Hades was following with him. And authority was given to them over a fourth of the earth, to kill with sword and with famine and with pestilence and by the wild beasts of the earth.
>
> (Revelation 6:8)

No matter what our individual religious convictions, the Four Horsemen of the Apocalypse continue to ride. We have engaged them in battle since the beginning of time. Are we so arrogant today that we honestly feel we have defeated them? We have not, for they return. The faces are different, but the same end is in mind. We have the tools to combat them. Let us not become so preoccupied with nonexistent threats that the next century becomes the final one.

What are the new weapons that humans have to combat these new forms of our timeless foes? The first is a safer, more selective, and more potent arsenal of drugs and chemicals that have been synthesized using both the complementary tools of organic chemistry and biotechnology. We have learned from our mistakes and have developed novel strategies to design specific chemicals that do not persist in the environment and kill their targets rather than the humans exposed to them. Next, we have learned more about the mechanisms of both disease and toxicity. We have sophisticated *in vitro* screening tests to identify compounds that actually eliminate a disease rather than just treat the symptoms. We can now screen for compounds that do not share the adverse properties of their ancestors, such as acute toxicity of the organophosphates, the persistence of DDT, or the estrogenic activities of some chlorinated chemicals. We have improved the safety of manufacturing and occupational use to further reduce side effects. All of these advances are well documented by substantial health statistics. We should move forward and not continually be preoccupied with nonexistent threats that modern toxicology has long ago silenced.

We are more environmentally conscious and have integrated ecological endpoints into the testing of new products so that they do not cause the problems of their predecessors. Toxicology continues to refine these endpoints to ensure that newly developed chemicals are as safe as possible. Finally, when mistakes do occur, we usually catch them earlier and develop remedial strategies. However, I believe we often overreact in this regard, when we associate detection of a chemical with actually causing an effect. Science can quantify the difference; it behooves us to educate the public to be able to interpret the conclusions.

We now realize that life is a complex web of interacting biological and ecological systems. Rachel Carson helped us appreciate this. However, we also appreciate that evaluating a single chemical in isolation of the others which it will come into contact with or using simple systems that do not reflect the complexity of organisms or populations is naive. We are developing strategies to address this integration. Extensive research is being conducted on how to make these extrapolations and how to factor in the complexity that is life. We are not yet there, but we continue to make phenomenal progress.

Biotechnology, molecular biology, and genomics are identifying the genes associated with diseases. This allows bioengineering solutions that include the now relatively straightforward synthesis of complex hormones (such as BST), the identification of gene protein products that cause disease, the designing of drugs to block their actions, and the use of gene-specific antisense drugs. Such drugs control the expression of genes rather than affect the activity of their protein products. Alternatively, one may actually insert new genes into the chromosomes of the target organism to change the phenotype so that the disease will not occur. Our bioengineered crops exemplify this approach. Finally, we are identifying the mechanisms and causes of aging and designing strategies to combat it. We are learning that the function of the immune system is central to both the defense against viral and bacterial invaders, as well as to the genesis of cancer and other diseases of senescence. These are future targets for both preventive and therapeutic strategies.

There are clearly ethical concerns to manipulation of the human genome. These must be addressed, taking into account all potential benefits and risks involved. These include the nature of the society that might result from such tinkering with the very molecules that define our species. The jury is out on these issues as the recent debates with stem cells and cloning clearly illustrate.

We are rediscovering that some aspects of our early ancestors' diets may actually be beneficial as the overwhelming data presented in this book about the benefits of eating fruits and vegetables should testify. We are now actively exploring what natural substances confer these benefits and devising ways to supplement the diet with them, or alternatively, bioengineer the plant to preferentially produce more by itself.

However, despite these tremendous advances in human health, science is being attacked. This is a continuation of a timeless conflict between different philosophical constructs of the world. As society is faced with an ever-increasing existential alienation and sense of social disintegration, we adopt other more positive world views, including mysticism and various resurrections of alternative sciences. If we can't control our destinies, then we create belief systems that make this come true. The paradox of modern technological man is that we can simultaneously adhere to conflicting scientific and anti-scientific world views. However, when the science behind our technological wonders is revealed and applied to other problems, we find that we do not truly understand the nature of science and technology.

Science is a process of learning about our world. It is only a world view to the extent that it demands that its tenets can be verified by experimentation. *It is not a value system.* Scientific data should only be interpreted in the context of the experiments that generated them, specifically in the context of the hypothesis that was formulated when the experiment was designed. As we saw with the Bendectin® example in the first chapter of this book, the same data can support diametrically opposed hypotheses. If an increase in the incidence of birth defects resulted from the use of this drug, this effect could be interpreted as either a drug effect or secondary to blocking the protective purgative effects of morning sickness! Drugs or chemicals are not inherently evil; their use must be judged relative to specific targets. Thalidomide, the drug which might have been responsible for the birth of modern toxicology due to its association with birth defects in the 1960s, was approved by FDA in 1998 to treat complications of leprosy and in 2000 has also been associated with effective treatment of proliferative disorders of the bone marrow.

One must always be sure that the endpoints of an experiment are compatible with the extrapolation to be made. When one assesses cancer in a laboratory or epidemiological study, one cannot use the data to make statements about other disease processes. One must assess whether the techniques employed are sensitive enough to show an effect. In other cases, one must assess whether the techniques are too sensitive and thus detect an effect that has no biological relevance. One must also decide if the statistics are sound:

- The study population is similar enough to the one that the inferences will be made about
- The exposure conditions are similar
- Appropriate doses are being used so that a dose-response extrapolation will be valid
- The study can account for sensitive populations

All of these issues have been addressed in this book. Science is a process and technique for learning about our world. It must be used properly if the results it generates can be applied to problems society deems important. This is where the abuse and confusion arise.

Science is fragmented into numerous small disciplines, each with their own arcane terminology and practices. The formulation of hypotheses that can be experimentally tested is central to all. *Science is a process, not an end. Science is not inherently good or evil. It is simply a tool to examine the world.* Value is only applied to science when it is used by humans. At the end of the last decade, society has stigmatized the very technology that allowed it to reach this phase of existence. It still is the best tool we have to combat many of the biological threats that will face us in the next millennium.

I am not a legal scholar; however, I see the legal system being based on historic precedence rather than new data from science. In science, hypotheses are meant to be disproven, thus the "truth" is never meant to be stable. I see the trend in law, awarding damages based upon an unfounded fear of disease rather than actual evidence of toxicity, as a serious problem facing society. We must educate the public that death is unavoidable, and one should not remove drugs or chemicals from the market based solely on grounds of emotion or compassion. The individual also has choices to make and must live by the consequences. Some behaviors, including diet and lifestyle choices, are risky to one's longevity. Society cannot be forced to bear the burden of care of diseases that are easily avoidable.

It is a paradox that we live in a high-tech society but are completely ignorant of its scientific underpinnings. Risk assessment, as presented in this book, involves both science and human judgement. It is the vagaries and vicissitudes of the latter that result in the problems we face today. The interface between science, law, and public policy will always be tenuous and fractious. Everyone involved in this process must attempt to understand, as much as possible, the underlying science so as to not take data out of context. This is an evolving discipline that must be allowed to use the latest science available and not be shackled to outdated laws and regulations we all know are no longer pertinent. Common sense must not be allowed to die.

> How extraordinary! The richest, longest-lived, best protected, most resourceful civilization, with the highest degree of insight into its own technology, is on its way to becoming the most frightened!
>
> (Aaron Wildavsky, 1979)

This prediction by political scientist Aaron Wildavsky has apparently come true. We have discussed in this book how anxiety caused by pseudoscientific threats may actually cause organic disease. Fear has taken hold and has become part of society's fabric. At the November, 1994, American Heart Association meeting in Dallas, some researchers reported that anxiety felt by patients over the fear of serious cardiac problems may actually trigger heart attacks. We saw with multiple chemical sensitivity, how associating certain symptoms with nondescript chemical exposure may condition this response to occur in some patients. This may even be facilitated by the olfactory/limbic system connections discussed previously. This strong psychological influence is really just an extension of the conditioned reflexes experienced by Pavlov's dogs and seen in B.F. Skinner's approach to behavioral psychology. The simplest and most efficient remedy to correct this misconception in the public's eye is to show all of the data and dispel the myths. We cannot let this false belief become "real" by codifying them into law. We must aggressively treat chemophobia by educating the public.

Toxicology faces new challenges as the products of biotechnology move from the laboratory to the clinic. A new toxicology must be formulated to take into account the different mechanisms of action these new products have. However, we must live in the present and anticipate the future, not be prisoners of the past. These molecules are significantly different from those developed in the middle of the last century. Just because a chemical kills an insect, and is thus classified as a pesticide, doesn't mean that it should perpetually be associated with DDT. EPA has promulgated new regulations on assessing cancer risk by compelling regulators to look beyond laboratory animal studies. Factors that should be taken into account include the specific mechanism of action of the chemical in question, the relevant pharmacokinetic data in both the laboratory animal and the human target, how the molecule compares with other known substances on the basis of a structure/analysis study, and the development of new transgenic animal models which better mimic the human response. These exciting approaches are better science and should result in better risk assessment. However, humans still will have to devise the regulations to use them, and the media will have to fairly report their conclusions.

This book should have convinced you that there is minimal risk and probable benefit from eating vegetables and fruits contaminated with trace levels of pesticides. There are some serious toxicological problems confronting us, many in the form of behaviors that are avoidable such as smoking and alcohol consumption. It is obscene to spend significant effort worrying about a part-per-billion exposure to a pesticide on an apple while drinking a cocktail and smoking a cigarette. In fact, eating that apple may be the best thing one can do to counteract the toxic effects of these habits if avoiding them is not possible.

I thus end where I began. This book is a plea to let old arguments die. Pesticide residues at today's low levels are not a human health concern. Modern science has developed extremely efficient approaches to reduce the toxicity and biological persistence of modern pesticides. DDT belongs in the history books, not the evening news. In the United States, we are dying from diseases associated with our prolonged life span. We have hard data to show that eating vegetables and fruits will prolong our life span and an affluent lifestyle. Chemophobia and stigmatization of biotechnology, and science in general, will not eliminate our problems. In fact, we have to muster all of our resources to combat the newly emerging diseases that truly threaten our existence.

Society's lack of discrimination between important and relatively harmless food safety issues has resulted in an unnatural obsession for protecting the public against minor, but easily measurable and thus pseudoquantifiable, threats. This is costly and does not significantly reduce actual hazards. The inability to distinguish serious hazards from those of little concern results in a confused populace that begins to ignore real threats to public health. Alternatively, a food and chemical phobia based in paranoia rather than true science evolves. Any real attempt at communicating risks becomes futile. Although we may feel assured that this approach will protect us from all harm, it will prevent us from enjoying the benefits of these products that, in the long run, may be our best approach for promoting health. Life does not come with a "money-back" guarantee. We often must accept small risks to avoid the more serious consequences that might affect us if we become complacent and forget that death and disease, for most, are a result of nature and not a creation of man.

There are social, political, ethical, and moral issues concerning quality of life, as well as the limits to which the human population should be allowed to grow, that must be addressed. These should be faced squarely and not cloaked in unrelated fears of pesticide-induced cancer epidemics. If one believes that population growth must be stopped before we exceed the carrying capacity of the earth, that technology should not be used to extend this carrying capacity, or that other byproducts of technology are destroying the ecosystem, then these must be debated and addressed on their own merits. The answer is not to ban pesticides on the false premise that they cause disease or "pull the plug" on agriculture so that limitations on food supply will stunt population. If one truly believes that smaller farms better support the social fabric of society and that the use of a drug, such as BST, works against this goal, then the argument should be made on those grounds and not on the false claim of a threat to human health. If one is morally opposed to genetic engineering, then this is the argument to be taken.

It was not the purpose of this book to solve these problems. Rather it was to illustrate how science can easily be misused when taken out of context. I'll end this book as I did the essay that prompted it:

> Well, I must go poison my children with apples and tomatoes (just kidding). I'll continue to buy fruits and vegetables from the grocery store since I know that in the long run, it's probably my kid's best chance to avoid cancer and serious health problems.

So stop worrying! Eat those veggies and drink that milk. Science and technology have greatly improved the safety of our food supply. *Bon Appetit!*

Toxicology Primer

Most of the material and debates presented in this book depend on knowledge of some areas of science, such as toxicology or chemistry, that are very
technical and well beyond the intended scope of this book. However, an
understanding of some basic concepts of toxicology is essential to assess the
safety of these chemicals. At high doses many pesticides are indeed poisons.
As fully explained in Chapter 2, dose makes all the difference. In addition,
each pesticide is an individual entity and may have unique properties you may
have come across in other readings. However, often these are not representative of all members of that class; the clearest example I used to show this was
DDT. Similarly, just because some pesticides, as discussed in Chapter 10, were
also used as chemical warfare agents , doesn't mean that your can of flea spray
can be used to terrorize New York! This level of "guilt by association" has
about as much validity as many of the arguments linking diverse groups of
chemicals under the single umbrella of "toxic pesticides."

This appendix serves as a brief technical supplement to the toxicology
introduced in the text. It is by no means intended to be comprehensive but
rather is designed to provide a more in-depth introduction to some technical
material covered in the text. Secondly, it introduces the reader to specific pesticides and their known toxicology. Critical concepts are defined where needed and the fundamentals of the toxicology of each class of pesticides are
introduced with as little technical jargon as I believe is safe. The interested
reader is urged to consult either listed textbooks or the citations from the original literature used to write this for further details. *My goal is to make available
some information on the process of setting tolerances and on high-dose and acute
pesticide toxicology so the main text of the book is not unduly burdened by technical*

details and complex regulatory jargon and procedure. I hope that this serves as a useful overview of this extensive and highly diverse discipline. Finally, a word of caution to the uninitiated reader of toxicology texts. Things are listed in toxicology texts because they do harm to living things. Chemicals that are harmless do not find their way into these texts.

The Establishment of Legal Tolerances

I would like to walk through the actual process of determining the safe levels of chemical exposure. There are many phases to this process. One is an experimental determination of *hazard* that is based on well-defined and well-monitored laboratory animal studies. The second is an estimate of *exposure* that involves how much residue is found in the food that is actually consumed by people. There are then subjective *safety factors*—some might call them "fudge" factors—added to protect against uncertainty and outright mistakes. Finally, the *risk* is calculated and thresholds set.

The first component we will discuss is the assessment of actual hazard, usually done with laboratory animal tests. Depending on the regulatory agency involved, studies are done in the most sensitive animal species using the most discriminating test available. Such safety studies are usually conducted in at least two rodent species, and any evidence of toxicity determined. The strains of mice and rats used are sometimes dependent on the type of chemical to be tested. In reality a number of sequential studies are conducted to determine the acute toxic dose and to define a profile of the disease induced. For some compounds or endpoints, such as birth defects, other animal species may be used, such as dogs, monkeys, or rabbits. Longer studies are then conducted to determine so-called *subchronic* effects and help select the doses used for chronic studies. The aim of these experiments is to determine the "no observed adverse effect level," the NOAEL.

The NOAEL is the dose at which *no* adverse effects are detected in the laboratory animals after chronic administration. It is the actual number determined by experimental studies in *sensitive* laboratory animal species dosed for long periods of time. The point of and meaning of the NOAEL must be stressed. In these studies, higher doses of compound will often elicit a toxic response. The design of the studies dictates that a dose must be selected which does not produce the response observed at higher doses. Thus these studies, usually conducted in two species of sensitive animals, determine a *safe* level of exposure.

Studies used to determine whether a chemical is carcinogenic, the so-called cancer "bioassay," are conducted in separate life-long studies. Toxicology data collected in the above trials are used to determine the maximum tolerated dose (MTD) of a compound. This dose, and one-half the MTD, are then given to animals for their entire lives to see if any cancer is detected. The MTD actually produces some adverse biological effect in the animals being treated. This is to ensure that the chemical is getting to the animal in a biologically active form. By definition, the MTD is greater than the NOAEL. It is at these higher doses that cancer is often detected in a single species and thus forces the chemical to be listed as a suspected carcinogen.

Many natural food constituents, tested at their MTDs, are often determined to be carcinogenic. This is the source of the controversy since they are identified as carcinogens by virtue of a positive result in a sensitive, adverse-effect producing, high-dose rodent bioassay. As discussed in earlier chapters, part of the problem of the MTD approach is that doses are selected which have overwhelmed the animal's natural defenses. There are many mechanisms that could result in such high levels of chemical exposure causing toxicity. *They are not present in an individual consuming a trace level of the same chemical in the diet!*

The actual design of these studies is determined by a combination of regulatory requirements and expert advisory committee input. These studies have to be conducted according to "good laboratory practices" (GLPs). These are an independent set of regulations designed to ensure that *all* aspects of the studies are done according to accepted, prereviewed, and *audited* protocols. Every phase of a study falls under the critical and painstakingly detailed requirements of this regulation. Chemicals must be of defined purity. Storage conditions are rigidly defined. Animals are acquired and maintained according to specific requirements. Their environment is monitored and recorded. Laboratory tests must be conducted precisely as outlined, using equipment that is documented, to be properly calibrated. All records must be kept in bound notebooks, and any entries must be signed by the individual recording them. Senior scientists and supervisors then must read and personally ensure that protocols were followed and the studies properly conducted. Random "on-site" quality assurance inspections and record audits are conducted to assure that all procedures are followed to the "letter of the law."

GLPs ensure that the data that are used in the regulatory process of setting tolerances are precise, defined, and not fraudulent. Many of these requirements were instituted when cases of sloppy, or even fabricated data, were exposed as being used in support of safety standards. These concerns are of *historical interest only* and have since been eliminated with widespread adoption of GLPs in the last three decades. Of course, one can never assure that conspiracy and fraud will be eliminated from all human activity. However, the health statistics reviewed in earlier chapters give assurance that not much has slipped through this regulatory gauntlet.

If a drug was positive in the MTD cancer bioassay described above, then the Delaney Clause often went into effect, and no amount of chemical exposure was considered safe. Under many of the regulations presently in force, the NOAEL is overruled and no further calculation is made. The level-of-detection dilemma comes into play. However, some agencies are starting to adopt approaches that recognize the problems inherent in the single molecule arguments. One example is how to incorporate data that a chemical is beneficial at very low doses due to hormesis. Agencies are using sophisticated dose-response models to calculate a so-called "benchmark dose" to replace the NOAEL. This often allows more knowledge of the chemical's properties, as well as other data, to supplement the more limited animal toxicology studies. Also, it is based on accepting some risk and extrapolating so that the risk of getting cancer from exposure to the chemical is less than one in one million. These approaches are constantly changing and are dependent upon the nature of the chemical, the regulatory agency involved and any specific legal statutes that might need to be considered.

The next step is to calculate the "acceptable daily intake" or ADI. If the chemical is not deemed to be carcinogenic, then the NOAEL is used. Alternatively, a dose such as the benchmark dose, based on dose-response modeling, may be used. One arrives at the ADI by dividing, and therefore making smaller, the NOAEL, by the uncertainty or safety factors. The regulatory agencies are responsible for establishing the ADI, and there are as many ways of doing this as there are regulatory agencies in existence. Some agencies, such as the EPA, may calculate a similar value, known as a reference dose (RfD). As we become more knowledgeable in these areas, the safety factors become more specific and, hopefully, the overall risk assessment more accurate.

As discussed in the bee analogy of Chapter 6, the minimal safety factor is usually 100x. This is determined by using a 10x factor for the uncertainty involved going from animals to humans and another 10x factor for different susceptibilities among humans, such as those in infants, children, the elderly, etc. If only these two factors are used, then the ADI equals the NOAEL divided by 100. In the bee example, this is what got us from a small room to a decent-sized lot. Other uncertainty factors may be used, especially if they cause birth defects. The EPA has instituted other 10-fold factors to account for deficiencies in study designs. These could include less than lifetime exposures if a NOAEL wasn't detected for a noncarcinogen, even if a minimally toxic dose was determined or if there were other indications of data defects. Every agency has a formula to estimate these risks, and often a factor less than 10 may be used if evidence suggests it is appropriate. For many food additives regulated by the FDA and USDA, a factor of 2,000 is often used. Other approaches are being adopted based on so-called "level of concern" with the chemical in question. This breaks chemicals into three groups of acceptable concentrations in food and assures that expensive toxicology studies are focused on situations that truly merit attention. The EPA hopes that by using its more sophisticated RfD approach, improved risk assessments can be achieved. For the EPA, the maximal uncertainty factor used is 10,000. Our fictional VIPA agency only used a 1,000-fold factor for bees since 10,000 would put the area at 1 bee per 33 acres!

This is a good time to highlight the fallacy in the warnings that say the setting of NOAEL in adult and healthy laboratory animal studies does not protect against the more susceptible infants and children. There are two points to counter these arguments. The first is that there is the 10x uncertainty factor that specifically is used to protect against a 10x increase in sensitivity. The recent National Academy of Science study of pesticides in the diet of infants and children came to this same conclusion. When they specifically addressed this issue, they concluded that the difference in susceptibility between children and adults is less than 10x. Therefore, the uncertainty factors already established in the system are adequate to protect infants and children. A problem with infants is that they are notorious for consuming a very restricted diet, such as only milk. Thus as we will discuss later, when the ADI is adjusted by food consumption factors to get at a tolerance in food, infant and adult eating patterns are different. Children often consume large percentages of processed food that may often reduce any level of pesticide that may have existed in the raw produce. If any risk was detected in this NAS study, it might be for acute poisoning from organophosphate insecticides, not chronic effects. Finally, recall that all of these safety factors assume that humans are *more* sensitive than laboratory animals. This is not necessarily true, and often animals may be more sensitive for a wide variety of reasons. This is a chemical-specific phenomenon. Thus for many chemicals, one could argue that this is even an additional biological safety factor.

The ADI or RfD is the measure of the *hazard* the chemical has for humans. It is that level of chemical that is deemed to be safe for daily lifetime human consumption. For "safe" residue levels to be set, *exposure* must now be determined. Every regulatory agency also has a method to accomplish this. The problem is determining how much food a person must eat to achieve this ADI. There have been numerous studies that have assessed the patterns of our food consumption to get at what is an average diet in our country. In cases such as milk, where the food may comprise the infant's entire diet, consumption is often overestimated. The presentation of these numbers is beyond the scope of this appendix since my purpose is only to illustrate the process that determines a pesticide *tolerance* in the field that is the "trip wire" for concern. This number is the concentration on the product itself, which based on patterns of food consumption, will determine the amount of daily consumption of the chemical.

Let us take a simple example to illustrate the process before we become buried in the arcane details of regulatory mathematics. Say that 1 milligram is the ADI for a chemical to be regulated. If an apple contained 1 milligram of this substance, only one apple could be consumed a day. The maximum tolerance allowed would be related to the concentration of chemical in apples that caused 1 milligram to be consumed if the entire apple were eaten. If an apple weighed 100 grams, the tolerance would be 0.01 milligrams per gram, or 10 milligrams per kilogram, which is equal to 10 parts per million. However, remember that this tolerance is based on the ADI that has at minimum a 100-fold safety factor already incorporated in it. In reality, 100 milligrams, or in this case 1000 parts per million of this chemical, was experimentally shown *not* to cause toxicity in the laboratory animals (100 apples/day in this scenario). Yet because of safety factors, 10 parts per million will be our apple benchmark.

This process for setting tolerances for drugs or other chemicals that may end up in meat or milk is somewhat easy to understand. The FDA and USDA establish these tolerances using the ADI and food consumption factors. These factors include how much of the food is consumed and what percentage of the diet it comprises. They then calculate a tolerance directly, a so-called maximum residue limit (you guessed it, there must be an acronym and it is MRL). This level cannot be exceeded when the drug is given to animals in the field. Milk and meat are tested to ensure that observed levels are well below these legal tissue tolerances.

Let us now specifically look at how the EPA sets pesticide tolerances for fruits and vegetables. First, recall that they use an RfD (what we have been calling ADI) to establish the safe level. The pesticide manufacturer submits a proposed level of pesticide in the field that would result if their product were used according to proposed label. This level is determined by field trials of pesticide usage and is the highest residue concentration that will likely be found based on good agricultural practices. These numbers are then factored into how much of the produce is consumed to produce a maximum residue consumption associated with eating crops treated with pesticides. If the food is processed and concentration occurs, this will be taken into account. An assessment is now also made of "aggregate exposure" to take into consideration potential human exposure from other sources besides food, such as home spraying, etc. The final step is to compare this maximum residue consumption with the ADI or RfD. For the purists and acronym lovers out there, this is termed the theoretical maximum residue contribution (TMRC). If it is lower than the ADI, then this is the approved *tolerance*. If it is not, then the manufacturer must lower application rates and approved dosages so the resulting tolerance will ultimately result in a daily exposure less than the ADI.

For our apple example, the manufacturer might suggest a pesticide that, when applied according to the label, produces a residue of 5 parts per million on the harvested apple. Since this is less than the 10 parts per million threshold that would exceed the ADI of 1 milligram per day, the approved tolerance would be set at 5 parts per million. This is the highest residue concentration likely to occur after agricultural usage of this product. Since it results in a daily intake less than the ADI, it is assumed to be safe. Thus in many cases the actual *tolerances* are even less than the safe levels. They reflect farming practices and are not a result of toxicology testing. However, they are approved based on how consumption would compare to the ADI, which is an estimate of toxicity.

All risk assessment ultimately boils down to establishing some level of chemical that should be presumed *safe*. Only when the levels *exceed* these thresholds should one become concerned.

A DIGRESSION ON THE CONCEPT AND LANGUAGE OF PERSISTENCE

Before any specific chemicals are reviewed, we should digress and discuss a property of certain compounds that is often associated with adverse effect. Its pertinence to many aspects of environmental toxicology will be obvious. To quantify both the biological and environmental persistence of chemicals or pesticides, scientists and mathematicians have developed a concept called the half-life ($T_{1/2}$) to rank and compare very different drug and chemical classes. A half-life is defined as the length of time it takes a chemical to degrade to one half or 50% of its original dose or amount applied. If I applied 100 lbs. of a pesticide to a field, the half-life is the time it would take for 50 pounds to degrade. If we talk about a drug or pesticide dosed in a human or animal, it is the time it would take for half the drug to be eliminated or destroyed by the body. One of the common denominators between many of the chemicals which have caused problems in the past is their tendency to have prolonged biological and environmental half-lives.

There is a peculiar logic associated with half-lives. In the above example, the half-life for our pesticide was 1 month. If we applied 100 lbs., we would only have 50 lbs. one month later. If we continue this, we would have 25 lbs. at the end of two months, 12.5 at the end of 3 months, and at the end of a year, we would be left with 0.0244140625 lbs., or 0.39 ozs. If we doubled the amount of pesticide put on a field, we do not double the length of time it takes to degrade it. We only increase it by another half-life. In our example, if we doubled the amount and follow the above halving procedure, it would only take 13 months to get to this same point. This is just one more half-life! If we applied 200 lbs., we would be back to 100 lbs. in 1 month, which was the starting point of the above exercise. The same logic holds for drugs in the body. If we doubled the exposure of anyone to a pesticide or drug, it only takes one half-life longer to get back to where we were at the lower dose or exposure. Half-lives are the *language* of the toxicologist, and if one wants to use "common sense" in this field, one must think in half-lives. Doubling the dose does *not* double the length of time a field, animal, or human is exposed to a chemical; it only increases it by one half-life. Hence compounds with short half-lives, even if given to animals or the environment at higher doses, do not produce proportionately prolonged exposure. Problems only occur when the half-lives are very long, like those seen with the chlorinated hydrocarbons, or at very high doses that overwhelm the body's ability to remove the substance. This is usually a problem of high-dose, and not low-level residue exposure.

For those readers more interested in this thought process, it is the reverse of population growth and doubling times. It is actually the mathematics of exponentials, but I promised to leave mathematics out of this book. Anyone who can divide by 2 or multiply by one-half can easily calculate these exposures.

I believe this parameter is so important because the difference in environmental and biological half-lives is the primary factor separating the pesticides and many environmental contaminants into "good" and "bad" groups. Those with long half-lives are persistent in the environment and in living organisms, and thus strike fear in the minds of chemophobics, since they never seem to leave the picture. Ironically, their persistence was the original reason they were developed, since they would not have to be applied as often in the field. However, this persistence can obviously generate problems since repeated application results in accumulation, especially for some compounds whose half-lives are measured in years rather than days or months.

CHLORINATED HYDROCARBON INSECTICIDES

The halogenated hydrocarbons of concern are classified as either chlorinated hydrocarbon insecticides or industrial chemicals and their by-products. The chlorinated hydrocarbon insecticides include DDT and its analogs, chlordane, heptachlor, aldrin, dieldrin, mirex, toxaphene, and lindane, among others. These are the compounds now banned in many countries and are also those still found to persist in environmental exposures. The industrial chemicals, and their by-products, include such substances as the polychlorinated biphenyls (PCBs), polybrominated biphenyls (PBBs), the chlorinated benzenes, the dioxins, and the dibenzofurans. Popular coverage of these compounds often lump them all into a single category because of the presence of a halogen (chlorine or bromine) or the *past* association of some as manufacturing contaminants When one chemical is condemned, all are condemned through guilt by association. Elimination of this lumping phenomenon by the media would go a long way in reducing the public's anxiety over pesticide exposure.

These compounds share certain physiochemical characteristics that favor their accumulation and persistence in animal and human tissues. All of the halogenated hydrocarbons are chemically stable and, to varying degrees, are resistant to degradation in the environment. They are lipophilic (fat soluble) and tend to accumulate in living organisms at concentrations greater than those in the environment. All show moderate to high resistance to metabolic transformation or metabolism by living organisms. They also have a low vapor pressure, which means they do not evaporate. This makes inhalational exposure minimal unless they are by-products of a combustion process such as incineration or smoking. This chemical stability and low volatility give them a prolonged persistence in the environment.

These properties give the halogenated hydrocarbons half-lives that generally range from a few weeks to several years and even decades, depending on the specific chemical entity and the animal species in question. Mathematical (pharmacokinetic) models developed to describe compounds with prolonged half-lives are usually complex involving two, three, or more components governing behavior at different concentration ranges.

The lipophilic nature of the halogenated hydrocarbons favors their partitioning from water to less polar solvents (e.g., fats, oils). Because cell membranes of animals contain substantial amounts of lipids, absorption of the chemical is facilitated. Once absorbed, the chemical penetrates into tissues with the highest neutral lipid content. Adipose tissue eventually makes up the primary reservoir, but there remains an equilibrium between the chemical in lipid-rich versus nonlipid-rich tissues. For animals with substantial amounts of body fat, this equilibrium greatly favors adipose tissue. For animals with very little body fat, equilibrium will be shifted toward leaner tissue. Unlike many other chemicals more subject to metabolism, often the only way to eliminate these compounds is by eliminating the fat. Because the direct excretion of neutral lipids is somewhat limited to sebaceous (oily sweat) secretions, egg yolks, and milk, the half-lives of these compounds are prolonged due to the low rates of excretion. The half-lives are shortened (increased excretion rate) in laying birds and lactating mammals since the chemicals are excreted in these products and passed on to offspring. Since milk is a primary route of excretion, one can easily understand the concern this raises for infants whose sole diet is milk.

Some members of this group do undergo metabolism. Metabolic conversion of the halogenated hydrocarbons to more polar (hence water-soluble) compounds greatly enhances their elimination. The rate at which this metabolic conversion occurs is inversely related to the half-life of the compound. The tendency for these compounds to be metabolized can be related to the number and position of the halogen atoms about the aromatic ring that comprises the molecule's backbone. The absence of two adjacent unhalogenated carbon atoms severely hinders the metabolism of the halogenated hydrocarbons. For compounds like hexachlorobenzene and mirex, there are no unsubstituted carbon atoms, and half-lives greater than 900 years have been estimated in some species! This translates into permanent residence since it exceeds the animal's life span.

Biotransformation of certain chlorinated hydrocarbon insecticides results in their conversion to metabolites which are less polar than the parent chemical. Heptachlor and aldrin are converted to the more lipophilic compounds heptachlor epoxide and dieldrin, respectively, whereas DDT is converted to DDE. The primary residue of DDT, which persists to the present day in animals and humans after exposure over a decade ago, is DDE. Following biotransformation, these compounds distribute to tissues which are higher in neutral lipid content than are the major organs of metabolism and excretion, the liver and kidney. These lipid-rich tissues are relatively, deficient in the so-called mixed-function oxidase (MFO) enzyme systems necessary for biotransformation of the halogenated hydrocarbons to more polar and thus more easily excreted compounds. As a result, these lipophilic chemicals remain unchanged in adipose tissue with only limited amounts returning to the circulation for possible metabolism and excretion. Paradoxically, aldrin and heptachlor metabolism results in an increased rather than reduced body load. This is opposite of the pattern seen for most other pesticide classes.

The classic prototype chlorinated hydrocarbon pesticide is DDT (dichlorodiphenyltrichloroethane), a chemical which earned its discoverer, Dr. Paul Müller, a Nobel Prize in 1948. This honor was an event that partly reflected the tremendous *beneficial* impact that this class of compounds had on disease prevention and food production. Although this compound is not used any longer in the United States (see Chapter 3), it is still widely used throughout the world where the risk/benefit ratio favors continued pesticide use. This is because of its efficacy in agriculture and its ability to kill insect vectors of important human diseases such as malaria. It is thus monitored in the FDA food surveillance programs.

The chlorinated hydrocarbons can be classified into four distinct chemical classes:

1. *Dichlorodiphenylethanes:* DDT, which includes DDD, DDE, and TDE; dicofol; chlorobenzilate; methoxychlor; and chloropropylate
2. *Hexachlorocyclohexane:* lindane; benzene hexachloride
3. *Cyclodienes:* aldrin; chlordane; dieldrin; endosulfan; endrin; heptachlor; photochlordane; toxaphene
4. *Miscellaneous* chemicals: mirex and chlordecone (kepone)

Hexachlorobenzene (HCB) and pentachlorophenol (PCP) have similar toxicological properties to the chlorinated hydrocarbons and often are discussed in the same breath.

As with any pesticide, there are acute and chronic manifestations of poisoning. A major manifestation of chlorinated hydrocarbon toxicity, which may have ecological implications, are the estrogenic and enzyme-inducing properties, which are especially pronounced with the dichlorodiphenylethanes. This interaction with steroid hormone metabolism at very high doses, such as that seen only after point contamination or environmental spills, may adversely affect reproduction. *This is not believed to be a concern at residual or trace levels encountered in the diet.* Although DDT and other members of this class have been weakly implicated to be carcinogenic in mice, they are not in the rat, dog, monkey, or hamster. Similarly, they are *not* listed as "official" human carcinogens. Therefore, even with the prototypical *evil* DDT, there is no hard evidence of serious human disease.

There are few recorded fatalities associated with the use of DDT-like pesticides, an attribute not shared with the other classes of chlorinated hydrocarbons, such as the cyclodienes. Most symptoms of acute dichlorodiphenylethane poisoning are related to oral ingestion (they are not well absorbed through the skin), and are manifested by signs related to the nervous system. An oral dose of 10 mg/kg may produce effects. For a "normal" 70 kg person, this is 700 mg or .7 of one gram! These signs include paresthesia (which is a prickling or tingling sensation), ataxia, dizziness, nausea, vomiting, fatigue, and peripheral tremors. In contrast, chronic signs are more diffuse and include weight loss, anorexia, anemia, tremors, muscle weakness, "nervous tension," anxiety, and hyperexcitability. These changes are associated with damage to the sensory arm of the peripheral nervous system. Pathology at these doses may also be seen in the liver and reproductive organs.

The cyclodienes are significantly more toxic after high-dose acute exposure and do direct damage to the central nervous system. Convulsions and seizures may be the first signs of acute toxicity. This is also associated with gastrointestinal symptoms, dizziness, headaches, general motor hyperexcitability, and hyperreflexia. Chronic signs are an extension of these and, in addition to muscle twitching, myoclonic jerking (muscle spasms), hyperexcitability, and headaches, there are psychological disorders and ultimately loss of consciousness and epileptic-like convulsions. In contrast to the DDT-like compounds, the cyclodienes are readily absorbed through the skin and thus are an occupational hazard to workers. This is the focus of toxicologic concern. Aldrin and dieldrin are also known reproductive toxicants at high doses. Mirex is a nervous system stimulant and also produces skin and ocular manifestations.

It must be recalled that lindane is a member of this class of compounds. It is used extensively in humans, including children, as an over-the-counter treatment for head lice. In veterinary medicine, it is widely used in sheep. When used at approved doses, this is an extremely safe product and has not been associated with any serious problems. The prolonged biological half-life associated with chlorinated hydrocarbons is shared by lindane, and thus results in long withdrawal times before sheep can be marketed for food.

Hexachlorobenzene and pentachlorophenol are chlorinated hydrocarbon fungicides that have been widely used in agriculture. HCB was associated with a classic case of mass poisoning in Turkey in the 1950s. Effects included photosensitization (sunlight-induced skin damage), hyperpigmentation (skin discoloring), alopecia (hair loss) and skin blistering. This compound is also a carcinogen as well as an immunosuppressant and alters porphyrin (a breakdown product of blood cells) metabolism. It is no longer produced in the United States. Pentachlorophenol, another chemical studied in my laboratory, is a widely used fungicide and wood preservative which is easily absorbed across skin. At very high doses, PCP has been shown to uncouple oxidative phosphorylation, resulting in hyperthermia, profuse sweating, dehydration, dyspnea, and ultimately coma and death. It has not been shown to be carcinogenic. One problem with PCP is that commercial exposures are often associated with chlorinated dibenzodioxins and dibenzofurans, as manufacturing contaminants, which produces signs of hepatic toxicity. Purified PCP alone is not believed to have as high a toxic potential although it has been reported to be fetotoxic. As with many of the pesticides, occupational exposure is the focus of work today, and the toxicology at these high levels of exposure *cannot* be compared to background levels.

This problem of manufacturing purity, which is now only of historic interest, was responsible for much of the adverse effects of past compounds. With today's improvements in manufacturing process chemistry, adoptions of rigid good manufacturing practices, and emphasis on quality control (with punitive legal action possible if procedures are not followed), *these issues are no longer relevant.* However, the limitations of these older studies, many using technical grade material that would not even be available today, are often forgotten when they are cited in relation to toxicologic concerns of related compounds.

ORGANOPHOSPHATE AND CARBAMATE INSECTICIDES

These classes of pesticides are among the most widely used and studied insecticides. They are defined based on their mechanism of activity, which is to inhibit the cholinesterase enzyme found in nerve junctions (synapses). Their major advantage over the chlorinated hydrocarbons insecticides are their *much shorter* biological and environmental half-lives, which are measured in hours and days rather than months or years. However, the *trade-off is that they have higher indices of acute toxicity.* Because of their short half-lives, tissue residues and chronic toxicity are generally not of concern.

There are a number of classifications of anticholinesterase insecticides that are primarily based on chemical structure. The organophosphates are chemically esters of either phosphoric (P=O) or phosphorothioc (P=S) acid. The carbamates are esters of carbamic acid. The various types of substituents groups to the phosphorus atom (alkyl, alkoxy, amido, aryl) further define the class and thus the chemical and toxicological properties which characterize them. The simplest are the *phosphates, phosphorothionates* and *thiolates*. Examples of phosphates are crotoxyphos (ciodrin), dichlorvos, paraoxon (parathion's active metabolite), and tetrachlorvinphos (rabon, gardona). The phosphorothionates and thiolates include chlorpyrifos, coumaphos, diazinon, famphur, fenthion, dichlofenthion, methylparathion, parathion, pirimiphos, and ronnel (trolene). The next large chemical class is the *phosphorothionothiolates*, which also include the *phosphorodithioates*. Examples of this class are carbophenothion, dimethoate (cygon), dioxathion, malathion, and phosalone. The *phosphoramides* include crufomate. The single-carbon phosphorus-substituent compounds include the *phosphonates*, such as trichlorfon (metrifonate); the *phosphonothionothiolates* such as fonophos; and the *phosphonothionates* such as EPN and leptophos. Carbamate insecticides include carbaryl, carbofuran, and propoxur.

The specific nature and composition of the substituents which define the class of compounds also determine the intrinsic potency against cholinesterase inhibition, the efficiency of absorption, and the pattern of metabolism seen in both target insects and mammals. It is misleading to talk of anticholinesterase pesticides as a single group since each compound has individual properties and activity against both insect targets and man. The common property is that these pesticides are very unstable in the environment and must be reapplied for effect or formulated in slow-release products.

The potential for acute toxicity can best be appreciated by examining the history of their development in the first half of this century. The first organophosphate pesticide, TEPP (tetraethylpyrophosphate), was developed before World War II by Dr. Gerhard Schrader in Germany as a replacement for nicotine. However, this compound was very unstable and toxic to mammals. The development, augmented by work in England, then diverged along two independent tracks—one for agricultural use, which focused on increased selectivity (higher insect/mammalian toxicity), and the second for chemical warfare use (see Chapter 10), focusing on significant mammalian toxicity. The first direction generated parathion, which became one of the most widely used pesticides in the world. The second effort yielded the chemical warfare nerve agents, such as soman (GD), tabun (GA), and sarin (GB), which are chemicals with intravenous human lethal doses as low as 10 micrograms per kilogram.

One must appreciate that all cholinesterase inhibitors cannot be described easily under a single umbrella since it would include the widely used carbamates found in most household pesticides along with the deadly chemical warfare agents, which are some of the most potent chemicals ever developed by man. Although generalizations are sometimes useful, in the field of risk assessment they can be misleading and sometimes blatantly wrong.

Regardless of their subclassification, all of these compounds have the identical mechanism of action, which is inhibition of acetylcholinesterase at nerve junctions where the molecule acetylcholine is the neuotransmitter. Most acute signs of toxicity are expressed as uncontrollable activity of the nervous system, which clinically is presented as salivation, lacrimation, urination, defecation, and dyspnea. After *lethal* doses, death results from failure of the respiratory system. Variations in the specific nerves affected, in how the body metabolizes the individual chemical, in where the chemical enters the body, and in the route of administration employed will change the specific clinical presentation seen for an individual exposure scenario.

Inhibition of acetylcholinesterase, the enzyme responsible for ending the transmission of a nerve impulse in our body, is due to a reversible change in its structure. A major variable in this process is that with continued exposure, or, for some specific chemicals, after any exposure, the organophosphate interaction with this enzyme may "age" and become permanent. In scientific lingo, this is a covalent nonreversible inhibition of the cholinesterase due to phosphorylation of its active site. This results in a prolongation and potentiation of toxic effect. This phenomenon is seen with the most toxic members of this class of organophosphates (for example, the chemical warfare agents) but *not* with the safer compounds. This is the primary difference between the carbamate and organophosphate classes since *the carbamates do not undergo this aging process* and thus their actions are readily reversible. This also significantly decreases their inherent toxicity relative to the organophosphates. For some compounds, "aging" can be reversed through the use of so-called regenerators such as pralidoxime chloride (2-PAM), one of the antidotes issued to our soldiers in the Gulf War in the event of chemical warfare attacks.

There are some other toxicologic manifestations of organophosphates that deserve mention. For example, accumulation of low levels of chemicals that irreversibly bind to cholinesterase may result in no signs of overt toxicity until the critical threshold is passed, whereby there is insufficient cholinesterase present to control synaptic function. This effect will *not* occur after ingestion of trace amounts in food. As discussed in this book, the scenario applies to the simultaneous home use of flea bombs, animal dips, carpet sprays, and no-pest strips in the same room. The signs are acute intoxication. This is a common presentation to veterinarians for dogs living in households of overzealous pet owners.

Other cholinesterases are also inhibited by these pesticides. For example, plasma aliesterases, which normally destroy circulating pesticides through ester hydrolysis, also are inhibited. Thus slow accumulation of one compound may bind to both target cholinesterases and detoxifying esterases. This is more pronounced with those agents that tightly bind to the enzyme. In such cases, exposure to normally nontoxic levels of a second organophosphate or carbamate could trigger acute toxicity. This exposure can be detected if plasma cholinesterase is assayed and depression is observed in the absence of clinical signs of neurotoxicity. Other manifestations of organophosphate-induced toxicity include blockage of acetylcholinestersase in the central nervous system where acetylcholine accumulation results in tension, anxiety, headache, psychological changes, and ultimately tremors, ataxia, convulsions, coma, and death due to respiratory failure. Again, these are *high-dose* effects.

There is a final toxicologic manifestation of organophosphates that is mediated by a different mechanism of action. This was originally described in 1930 as "ginger paralysis" or "jake leg," secondary to a Jamaican rum additive contaminated with the organophosphate triorthocresyl phosphate (TOCP). The syndrome, now termed organophosphate-induced delayed neuropathy (OPIDN), is a neurotoxicity seen 10 to 14 days post chemical exposure associated with motor weakness and sensory loss. Severely affected patients are ataxic (lose control of muscles) and spastic. The pathology is characterized by a "dying back" and demyelination of neurons. The precise mechanism is under intense investigation. Permanent injury usually results. OPIDN has *only* been associated in humans with *high-dose* occupational exposure to DFP, leptofos, mipofox, merpos, and trichlorfon. This effect is not seen with other organophosphates and is *irrelevant* to the issue of low-level residues in food.

The treatment of organophosphate and carbamate intoxication is dictated by their common mechanism of action directed at cholinesterase inhibition. First, respiration must be maintained by artificial means since this is usually the cause of death. Second, atropine sulfate is given to competitively inhibit acetylcholine binding to the postsynaptic receptors. This effectively reverses symptomology. Finally, enzyme regenerators such as pralidoxime hydrochloride (2-PAM), pralidoxime methanesulphonate, obidoxime, or pyrimidoxime, should be immediately administered to help generate free enzyme by reversing the "aging" process discussed above. Diazepam is used to counter convulsions if present.

Most organophosphates are readily absorbed by oral, inhalational, and dermal routes of exposure. Acute toxicity is generally related to either accidental or deliberate ingestion or to occupational exposure. *It is accepted that low-level exposure on foods and in the environment is not sufficient to cause human health signs of nervous system toxicity such as described above. This is true even* for exposure to compounds with non-anticholinesterase effects. For example, many organophosphates are immunotoxic but only at doses which also produce neurotoxicity. The only exception noted to this is with the contaminant O,O,S-trimethylphosphorthioate (O,O,S-TMP), which causes immunosuppression in the absence of any other biochemical or biological effect. Some members of this group have been shown to be teratogenic at high doses, especially in avian embryo studies. Finally, some (phosphorothiolates) have been shown to be direct pulmonary toxicants; however, these compounds are not widely used as insecticides. These compounds have not been reported to have estrogen-like activities.

Toxicity can occur secondary to exposure to treated fields since many of these compounds may be easily absorbed across the skin. The potential for dermal absorption is compound dependent and varies from 2–70% of the applied dose. These compounds may also be metabolized while being absorbed across the skin, and the environmental conditions during exposure (temperature, relative humidity) drastically modify absorption. Similarly, solvent effects and coapplication of other pesticides may modify the amount absorbed, making risk assessment from single-chemical data difficult. This is a primary reason that occupational exposure, and not food residues, should be the primary focus of pesticide toxicology.

Miscellaneous Insecticides

This section includes the pyrethroid insecticides, rotenone, and some pesticide synergists and additives. There are also insect hormones which target physiological processes not present in mammals. The pyrethroids include the naturally occurring pyrethrins, and two classes of synthetic pyrethroids. The first class is known as type I compounds, which lack an α-cyano substituent. An example of a type I compound is permethrin. Type II compounds, which have an α-cyano substituent, include cypermethrin, fenvalerate, and flucythrinate. *These compounds are among the safest class of pesticides available.* The natural pyrethrin is an extract from chrysanthemum flowers. It was originally discovered by the Chinese in the first century A.D. and used in Europe in the 1800s. Toxicity associated with Type I synthetics is characterized by tremors that are mediated by a mechanism of action similar to that of the chlorinated hydrocarbon, DDT, and involves both central and peripheral effects. The second group is characterized by induction of clonic seizures, writhing, and salivation in rats, effects which are mediated by a different mechanism that may possibly involve binding to a specific receptor, GABA, limited to the central nervous system. Despite these signs, high doses are required to elicit an effect, making them far safer than organophosphate or chlorinated hydrocarbon insecticides.

Exposure to the natural pyrethrum may cause contact dermatitis and other allergic signs not seen with the synthetic compounds. However, some Type II pyrethroids may induce a cutaneous paresthesia lasting up to 18 hours. Pyrethroids do not accumulate in the body and are not associated with chronic toxicity or tissue residue concerns. They are easily susceptible to metabolism in many tissue sites by esterases and microsomal oxidases. They are also environmentally labile and only pose ecological problems to some aquatic species because of the extreme sensitivity of those species to the acute toxic effects of those compounds.

Rotenone is an alkaloid botanical pesticide isolated from plants (*Derris sp.* or *Lonchocarpus sp.*). It blocks mitochondrial electron transport. It is associated with dermatitis and mucous membrane irritation in humans and is very potent in fish. In humans, intoxication is rare but when present is directed toward the respiratory system. Rotenone is used as a topical ectoparasiticide. As mentioned in the text, it has been implicated as possibly having a role in Parkinson's disease. The newest and by far safest class of insecticides available today are the insect growth inhibitors, such as methoprene (PreCor®), and chitin synthesis inhibitors, such as lufenuron (Program®).

HERBICIDES

Herbicides are chemicals specifically designed to kill plants, thus they generally have a low order of toxicity for nontarget animals and humans. The focus of modern herbicide development is to make compounds that are selectively targeted to unwanted vegetation (e.g., weeds) while sparing desirable plants (e.g., crops, ornamentals). These chemicals are the most rapidly growing sector of the agrochemical business and one of its members, atrazine, is the highest volume pesticide used in American agriculture.

There are numerous ways these chemicals may be classified, including chemical structure, time of field application relative to expected growth of target plant (e.g., preemergent, postemergent), or plant-specific activity. Like the problems encountered with some pesticides, earlier studies often attributed significant mammalian and even human toxicity to specific herbicides. It was later discovered that this toxicity was due to the *presence of manufacturing contaminants*. The classic example often quoted in the popular press was TCDD (2,3,7,8-tetrachloro-dibenzo-p-dioxin) contamination of the herbicide Agent Orange (a mixture of 2,4-D and 2,4,5-T) in Vietnam and its alleged responsibility for all of the toxicologic manifestations associated with its use. TCDD was present as a contaminant due to early manufacturing processes that have since been corrected. One must examine the toxicology literature on these chemicals with care and ascertain exactly what form of herbicide was being reported. Finally, like many other problems encountered in this text, it is not a problem with use of today's herbicides.

The first herbicides that will be reviewed are the chlorophenoxy compounds, which include 2,4-D (2,4-dichlorophenoxyacetic acid), 2,4,5-T (2,4,5-trichlorophenoxyacetic acid), MCPA (4-chloro-2-methylphenoxy-acetic acid), MCPP (mecoprop, (±)-2-(4-chloro-2-methylphenoxy)-propanoic acid), and silvex (fenoprop; 2-(2,4,5-trichlorophenoxy)propionic acid). Silvex is an irritating member of this class which was banned by the EPA in 1985 for most uses in the United States. These compounds are plant regulators called auxins. They are phytohormones that destroy the host plant by inducing abnormal growth that disrupts nutrient transport. Mammals, including humans, do not regulate their metabolism using these phytohormones. Only plants do.

The most widely used members of this class of herbicides are 2,4-D and 2,4,5-T. Their low level of mammalian toxicity is evidenced by their LD_{50} values, which range from 0.3 to 1.2 gram per kilogram in all species studied. In a 70 kilogram person, this ranges from 21 to 84 grams of chemical. Exposure to lower sublethal doses of 2,4-D results in muscle-related symptoms such as myotonia. Neither compound is carcinogenic in animal studies. There also appears to be no evidence of cumulative toxicity. Occupational and accidental exposure studies in man generally report mild symptomology associated with neurological and muscular systems at oral doses greater than 50 mg/kg. There is little documented evidence of direct neurotoxicity. The residue levels found in food are completely outside of the ranges that produce these mild manifestations.

The most controversial aspect of their toxicology relates to the teratogenic potential of 2,4,5-T. This concern originally stemmed from the Agent Orange controversy. In some but not all studies, cleft palate and cystic kidneys have been observed at maternal doses as low as 15 mg/kg/day. Many of these studies were confounded by the presence of dioxin as a contaminant. Similarly, chloracne associated with occupational exposure to herbicides was determined to be a dioxin effect and no longer is a concern. These were high-dose exposures to workers manufacturing the chemicals. There is general agreement that human health concerns over 2,4-D and 2,4,5-T are vanishingly small if the dioxin issue is handled separately.

The next major class of widely used herbicides is made up of the bipyridium compounds, which include paraquat (1,1'-dimethyl-4,4'-bipyridylium dichloride) and diquat (1,1'-ethylene-2-2'-bipyridylium dibromide). Experimental exposure of animals, or accidental or deliberate exposure of humans to high doses of paraquat, produces respiratory toxicity after 10 to 14 days. Although this polar compound is poorly absorbed after oral delivery, it is actively concentrated in the lung. There it results in necrosis of pulmonary tissue followed by a proliferative phase that ultimately ends in pulmonary fibrosis and end-stage lung disease. The initial toxicity is directed at membranes and is a result of free radical oxidative damage secondary to generation of hydrogen peroxide and the superoxide anion. Large doses also result in membrane damage in other organ systems, with the kidney accumulating the drug. Diquat has a slightly lower toxic potential than paraquat, although diquat does not preferentially accumulate in the lungs. Thus pulmonary toxicity is not seen. Instead the gastrointestinal tract, liver, and kidney are affected. Because of the fixed charge of both these compounds, topical application does not result in toxicologically significant absorption, although some degree of skin damage may occur. There is no concern with these chemicals at low-level food exposure.

The acetanilides are another widely used set of herbicides and include alachlor, metolachlor, propachlor, and propanil. These compounds are extremely safe, having oral LD_{50} values of up to 10 g/kg/day in laboratory animal studies. In a 70 kilogram person, that would mean eating over a pound per day. Problems in humans are restricted to occupational or manufacturing exposure. The signs most often seen are anorexia and depression. High-dose animal studies of alachlor in rodents demonstrated the formation of nasal and stomach tumors. Propanil has been associated with methaemoglobinaemia formation, a general property of many aniline derivatives. However, the LD_{50} for propanil is 1.4 g/kg. Thus, these compounds are generally considered "safe" by toxicological standards, especially when compared to other categories of chemicals tabulated in this text.

The triazines make up the final general class of herbicides. This class includes the most prevalent pesticide in use for over two decades, *atrazine*. It also includes prometon, propazine, and simazine. Acute poisoning with the triazines results in anorexia, depression, and muscular symptomology. The oral doses in rodents required for toxic effects are greater than 1.74 g/kg, again making these compounds extremely safe. There is no evidence of toxicity from field usage reports.

There are a number of organic acids that interfere with plant metabolism. Dicamba (3,6-dichloro-2-methoxybenzoic acid) and tricamba are herbicides which interfere with protein synthesis, while picloram (4-amino-3,5,6-trichloro-2-pyriidinecarboxylic acid) interacts with nucleic acid synthesis in susceptible plants. Dicamba is primarily eliminated in the urine due to its hydrophilic properties. Some benzonitrile derivatives are also employed as herbicides, including dichlobenil (2,6-dichlorobenzonitrile), bromoxynil (3,5-dibromo-4-hydroxybenzonitrile), and ioxynil. Linuron (N'-(3,4-dichlorophenyl)-N-methoxy-N-methylurea) is a phenyl urea while trifluralin (a,a,a-trifluro-2,6-dinitro-N,N-dipropyl-p-toluidine) is a dinitroaniline. The major concern with the substituted ureas is induction of hepatic microsomal enzymes which could modulate the disposition of simultaneously exposed chemicals metabolized by the same enzymes. Finally, one of the most common herbicides presently in use is glyphosphate, or Roundup®, a phosphonomethyl amino acid, which was encountered in the text in relation to plants bioengineered to be resistant to this broad spectrum compound. This compound is safe in concentrations applied in the field, although ingestion of the concentrated solution has been recently used as a suicidal agent. Occupational exposure to high concentrations may result in a severe contact dermatitis.

Selected Readings
and Notes

INTRODUCTION

These are my two original works on this subject that serve as the foundation of the present book:

J. E. Riviere. "Stop Worrying and Eat Your Salad. Science and Technology Have Improved the Safety of Our Food Supply." In "My Turn." *Newsweek*, pg. 8, Aug. 8, 1994.

J.E. Riviere. *Why Our Food Is Safer Through Science: Fallacies of the Chemical Threat*. Fuquay-Varina, NC: Research Triangle Press, 1997.

The following are general books that discuss public overreactions to chemical threats and other issues:

J. Best. *Damned Lies and Statistics: Untangling Numbers from the Media, Politicians, and Activists*. Berkeley: University of California Press, 2001.

P.K. Howard. *The Death of Common Sense: How Law is Suffocating America*. New York: Random House, 1994.

D. Murray, J. Schwartz, S.R. Lichter. *It Ain't Necessarily So: How Media Make and Unmake The Scientific Picture of Reality*. Lanham: Rowman and Littlefield Publishers, 2001.

National Institute of Environmental Health Sciences. *Human Health and the Environment. Some Research Needs*. Report of the Third Task Force for

Research Planning in Environmental Health Science. Washington, D.C.: National Institutes of Health, (NIH Publication No. 86-1277), 1984.

The "Objective Scientists Model" as published in the 1995 organizational brochure of *The American College of Forensic Examiners*, 1658 South Cobblestone Court, Springfield, MO. (Note: This is similar to oaths and principles espoused by numerous scientific societies that oblige its members to be objective when evaluating and interpreting scientific data.)

CHAPTER 1

The first four research manuscripts listed here depict research that probably set this author's mindset that sensitive populations exist and must be accounted for in any controlled experiment:

J.E. Riviere. "A Possible Mechanism for Increased Susceptibility to Aminoglycoside Nephrotoxicity in Chronic Renal Disease." *New England Journal of Medicine*, Vol. 307, pgs. 252–253, 1982.

D.L. Frazier, L.P. Dix, K.F Bowman, C.A. Thompson, J.E. Riviere. "Increased Gentamicin Nephrotoxicity in Normal and Diseased Dogs Administered Identical Serum Drug Concentration Profiles: Increased Sensitivity in Subclinical Renal Dysfunction." *Journal of Pharmacology and Experimental Therapeutics*, Vol. 239, pgs. 946–951, 1986.

J.E. Riviere, L.P. Dix, M.P. Carver, D.L. Frazier. "Identification of a Subgroup of Sprague-Dawley Rats Highly Sensitive to Drug-induced Renal Toxicity." *Fundamental and Applied Toxicology*, Vol. 7, pgs.126–131, 1986.

D. L. Frazier, J.E. Riviere. "Gentamicin Dosing Strategies for Dogs with Subclinical Renal Dysfunction." *Antimicrobial Agents and Chemotherapy*, Vol.31, pgs.1929–1934, 1987.

K.R. Foster, D.E. Bernstein, P.W. Huber, eds. *Phantom Risk. Scientific Inference and the Law*. Cambridge, MA: The MIT Press, 1993. (Note: This is an excellent book illustrating how our present legal system is formulating judgments that are contrary to the principles of modern science. As discussed throughout my book, this phenomenon increases the public's belief in "pseudoscientific" explanations and increases anxiety even if there is no real biological or medical problem. The legal system then awards damages based upon a chemical's ability to cause anxiety, even if the underlying cause for this anxiety is false.)

R. M. Nesse, G.C. Williams. *Why We Get Sick: The New Science of Darwinian Medicine*. New York: Times Books, 1994. (Note: Chapter 6 of this provocative book gives an interesting perspective on natural toxins and raises the question: Are we any worse off substituting synthetic poisons for the old natural ones? As a note, Bendictin® is also presented and a hypothesis developed that morning sickness in pregnancy is actually good since it allows the mother to decrease her intake of natural toxins. Thus blocking it using Bendictin® would by itself increase the potential for birth defects. As can be seen by the data presented in the present book, this hypothesis would not be substantiated as birth defects decreased.)

J. Best. *Damned Lies and Statistics: Untangling Numbers from the Media, Politicians, and Activists.* Berkeley: University of California Press, 2001. (Note: This short but informative book also provides many illustrations of how statistics are misused and "mutated" to make specific points in the social sciences.)

Food and Drug Administration. "Indication for Bendectin Narrowed." *FDA Drug Bulletin.* Vol. 11, Number 1, pgs. 1–2, March, 1981.

J.F. Cordero, G.P. Oakley, F. Greenberg, L.M. James. "Is Bendectin a Teratogen?" *Journal of the American Medical Association.* Vol. 245, pgs. 2307–2310, 1981.

L.B. Holmes. "Teratogen Update: Bendectin." *Teratology,* Vol. 27, pgs. 277–281, 1983.

A.A. Mitchell, L. Rosenberg, S. Shapiro, D. Slone. "Birth Defects Related to Bendectin Use in Pregnancy. I. Oral Clefts and Cardiac Defects." *Journal of the American Medical Association.* Vol 245, pgs. 2311–2314, 1981.

G. Taubes. "Epidemiology Faces Its Limits." *Science,* Vol. 269, pgs. 164–169, 1995. (Note: This is a fascinating critique of epidemiology by its own practitioners which should be required reading material for all jurors, reporters, and anyone else who quotes the results of an epidemiological study).

CHAPTER 2

U.S. Congress, Office of Technology Assessment. *Pesticide Residues in Food: Technologies for Detection.* OTA-F-398, Washnigton, D.C.: U.S. Government Printing Office, October, 1988.

U.S. Bureau of the Census. *The American Almanac 1994–1995: Statistical Abstract of the United States, 114th Edition.* Austin: The Reference Press, 1994.

Center for Risk Analysis. *A Historical Perspective on Risk Assessment in the Federal Government.* Boston: Harvard School of Public Health, March, 1994.

P.W. Anderson. "More Is Different." *Science,* 1972.

B. Ballantyne, T. Marrs, P. Turner. *General and Applied Toxicology.* New York: Stockton Press, 1993.

A.L. Craigmill, S.F. Sundlof, J.E. Riviere *Handbook of Comparative Pharmacokinetics and Residues of Veterinary Therapeutic Drugs.* Boca Raton, FL: CRC Press, 1994.

W.B. Deichman, D. Henschler, B. Holmstedt, G. Keil. "What Is There That Is Not a Poison? A Study of the Third Defense by Paracelsus." *Archives of Toxicology.* Vol. 58, pgs 207–213, 1986.

M. Friedman, ed. *Nutritional and Toxicological Aspects of Food Safety.* New York: Plenum Press, 1988.

K.R. Foster, D.E. Bernstein, P.W. Huber, eds. *Phantom Risk. Scientific Inference and the Law.* Cambridge, MA: The MIT Press, 1993 (Note: Dr. Ames's estimate came from this source. See original articles in reading list for Chapter 5).

A.W. Hayes, ed. *Principles and Methods of Toxicology*, Third Edition. New York: Raven Press, 1994.

J. Horgan. "From Complexity to Perplexity." *Scientific American*, Vol. 272, pgs. 104–109, 1995.

J.E. Riviere, A.L. Craigmill, S.F. Sundlof. "The Food Animal Residue Avoidance Databank (FARAD): An Automated Pharmacologic Databank for Drug and Chemical Residue Avoidance." *Journal of Food Protection*, Vol. 49, pgs 826–830, 1986.

J.E. Riviere. "Pharmacologic Principles of Residue Avoidance for the Practitioner." *Journal of the American Veterinary Medical Association*, Vol 198, pgs. 809–816, 1991.

J.E. Riviere, A.L. Craigmill, S.F. Sundlof. *Handbook of Comparative Pharmacokinetics and Residues of Veterinary Antimicrobials.* Boca Raton, FL: CRC Press, 1991.

R. Stone. "Environmental Estrogens Stir Debate." *Science*, Vol. 265, pgs. 308–310, 1994.

S. F. Sundlof, J.E. Riviere, A.L. Craigmill. *Handbook of Comparative Pharmacokinetics and Residues of Pesticides and Environmental Contaminants in Animals.* Boca Raton, FL: CRC Press, 1995.

CHAPTER 3

D.T. Avery. *Saving the Planet with Pesticides and Plastic* Indianapolis: Hudson Institute, 1995.

J.L. Bast, P.J. Hill, R.C. Rue. *Eco-Sanity: A Common Sense Guide to Environmentalism.* Lanham, MD: Madison Books, 1994.

R. Carson. *Silent Spring.* Boston: Houghton Mifflin Co., 1962. (Quotation is from opening chapter on pg. 2.)

D.J. Ecobichon. "Pesticide Use in Developing Countries." *Toxicology*, 160: 27–33, 2001.

E. Efron. *The Apocalyptics: How Environmental Politics Controls What We Know About Cancer.* New York: Simon and Schuster, 1984.

A. W. Hayes, ed. *Principles and Methods of Toxicology*, Third Edition. New York: Raven Press, 1994.

J. Kaiser, M. Enserink. "Treaty Takes a POP at the Dirty Dozen." *Science*, 290: 2053, 2000.

L. Lawson. *Staying Well in a Toxic World. Understanding Environmental Illness, Multiple Chemical Sensitivities, Chemical Injuries and Sick Building Syndrome.* Chicago: The Noble Press, 1993.

N. Neil, T. Malmfors, P. Slovic. "Intuitive Toxicology: Expert and Lay Judgements of Chemical Risks." *Toxicologic Pathology*, Vol. 22, pgs 198–201, 1994.

D.L. Ray. *Trashing the Planet: How Science Can Help Deal with Acid Rain, Depletion of the Ozone, and Nuclear Waste (Among Other Things).* New York: HarperCollins, 1990.

F. Setterberg, L. Shavelson. *Toxic Nation: The Fight to Save Our Communities from Chemical Contamination.* New York: John Wiley & Sons, 1993.

J. Sherma. "Pesticide Residue Analysis (1999–2000): A Review." *Journal of the AOAC International.* 84: 1303–1312, 2001.

S.F. Sundlof, J. E. Riviere, A.L. Craigmill. *Handbook of Comparative Pharmacokinetics and Residues of Pesticides and Environmental Contaminants in Animals.* Boca Raton, FL: CRC Press, 1995.

R. Thurow. "As a Tropical Scourge Makes a Comeback, So, Too, Does DDT." *Wall Street Journal,* July 26, 2001.

Pesticide levels in produce tabulated in Table 3.1 were compiled from: Food and Drug Administration. *Pesticide Program: Residue Monitoring.* Thirteenth Annual Report covering October 1998 through September, 1999, Washington, DC.

Produce utilization values are from the U.S. Bureau of the Census. *Statistical Abstract of the United States: 2000, 120th Edition.* Washington, DC, 2000.

A.L. Aspelin. *Pesticide Industry Sales and Usage; 1992 and 1993 Market Estimates.* Washington, D.C.: Office of Pesticide Programs, Environmental Protection Agency (Publication 733-K-94-01), June, 1994.

K. R. Foster, D.E. Bernstein, P.W. Huber, eds. *Phantom Risk. Scientific Inference and the Law.* Cambridge, MA: The MIT Press, 1993. (See the chapter by Dr. Bruce Ames on dangers of natural toxins in hybrid potatoes and celery, pgs. 176–177.)

L. Helmuth. "Pesticide Causes Parkinson's in Rats." *Science*, 290: 1068, 2000.

J.E. Keil, S.T. Caldwell, C. B. Loadholt. *Pesticide Usage Survey of Agricultural, Governmental, and Industrial Sectors in the United States, 1974.* Charleston, S.C.: Epidemiological Studies Program Center, Report for Environmental Protection Agency, Office of Pesticide Program Contract 68-01-1950, June, 1977. (Data used to compile Table 3.2.)

D. Pimentel. *CRC Handbook of Pest Management in Agriculture, 2nd Ed., Volumes I, II and III.* Boca Raton, FL: CRC Press, 1991. (Used to construct Table 3.3.)

CHAPTER 4

All of the vital statistical data used in the chapter is from the latest edition of a compilation of census data (*Statistical Abstract of the United States: 2000, 120th Edition*). Most survey sources included are current through 1998. This was selected because it is unbiased and is the source of the data from which most analyses ultimately originate. This book is available in many book stores and on the web (*www.census.gov/statab*).

U.S. Census Bureau *Statistical Abstract of the United States: 2000, 120th Edition*. Washington, DC, 2000.

S. J. Olshansky, B.A. Carnes, A. Desesquelles. "Prospects for Human Longevity." *Science*, 291: 1491–1492, 2001. (Note: This argues for the 85-year limit to mean life expectancy unless all risks of death after 85 are removed.)

R. Carson. *Silent Spring*. Boston: Houghton Mifflin Co., 1962. (Quotation on cancer is from Chapter 14, pg. 221; that of liver disease is from Chapter 12, pg. 192.)

National Cancer Institute. *1987 Annual Cancer Statistics Review Including Cancer Trends 1950–1985*. Washington, DC: National Institutes of Health (NIH Publication No. 88-2789), 1988.

D. Murray, J. Schwartz, S. Lichter. *It Ain't Necessarily So*. New York: Rowman and Littlefield, 2001. (Note: This is an excellent reference which illustrates many pitfalls of statistics when applied to demographics and other issues.)

P. Schulze, J. Mealy. "Population Growth, Technology and Tricky Graphs." *American Scientist* 89: 209–211.

U. Hoffrage, S. Lindsey, R. Hertwig, G. Gigerenzer. "Communicating Statistical Information. *Science*. 290: 2261–2262, 2000. (Note: This short article stresses the importance of expressing statistical incidences as frequencies rather than percentages, a practice especially important when tabulating death rates since the number of possibly affected people always decreases at advanced ages.)

There are entire textbooks devoted to the effects of smoking and alcohol on chemical actions in the body. A few are listed here:

I.H. Stockley. *Drug Interactions*, 2nd. Ed., Oxford: Blackwell *Science*, 1991.

M. Gibaldi. *Biopharmaceutics and Clinical Pharmacokinetics*, 4th Ed. Philadelphia: Lea and Febiger, 1991.

R. Bailey. *Eco-Scam: The False Prophets of Ecological Apocalypse*. New York: St. Martin's Press, 1993. (Source of worldwide statistics.)

D.T. Avery. *Saving the Planet with Pesticides and Plastic*. Indianapolis: Hudson Institute, 1995. (Note: Breast cancer statistics were discussed on pg. 395;

Dr. Scheuplein's quote on the incidence of food-cancer is discussed on pg. 73, supporting Bast et al. sources below.) Some of this data was corroborated in the following reference.

C.H. Rubin, C.A. Burnett, W.E. Halperin, P.J. Seligman. "Occupation as a Risk Identifier for Breast Cancer." *American Journal of Public Health*, Vol. 83, pgs. 1311–1315, 1993.

J.L. Bast, P.J. Hill, R.C. Rue. *Eco-Sanity: A Common Sense Guide to Environmentalism.* Lanham, MD: Madison Books, 1994. (Note: The often quoted incidence of food-related cancer attributed to Dr. Scheuplein of the FDA is cited on pgs. 21–22.)

R. Doll, R. Peto. *The Causes of Cancer.* Oxford: Oxford University Press, 1981.

D. Waltner-Toews. *Food, Sex and Salmonella. The Risks of Environmental Intimacy.* Toronto: New Canada Press, Ltd., 1992.

S. Begley. "Beyond Vitamins." *Newsweek*, April, 25, 1994, pgs. 45–49.

S. Begley with D. Glick. "The Estrogen Complex." *Newsweek*, March 21, 1994.

S. Briggs and the Rachel Carson Council. *Basic Guide to Pesticides: Their Characteristics and Hazards.* Washington, D.C.: Taylor and Francis, 1992. (Note: Quote that vegetarians are known to have less cancer appears in "Appendix Three: The Meaning of Carcinogenecity Testing," by William Lijinsky, pg. 265.)

The following six references are an excellent illustration of the nature of the scientific inquiries into trying to find the active principle responsible for the known benefit of vegetable consumption on protection against lung cancer:

P.H. Abelson. "Editorial: Adequate Supplies of Fruits and Vegetables." *Science*, Vol. 266, pg. 1303, 1994.

G.A. Colditz, M.J. Stampfer, W.C. Willett. "Diet and Lung Cancer. A Review of the Epidemiologic Evidence in Humans." *Archives of Internal Medicine*, Vol. 147, pgs. 157–160, 1987.

E.H. Fontham. "Protective Dietary Factors and Lung Cancer." *International Journal of Epidemiology*, Vol. 19 (Supplement 1), pgs. S32–S42, 1990.

L.Le Marchand, J.H. Hankin, L.N. Kolonel, G.R. Beecher, L.R. Wilkens, L.P. Zhao. "Intake of Specific Carotenoids and Lung Cancer Risk." *Cancer Epidemiology, Biomarkers and Prevention*, Vol. 2, pgs. 183–187, 1993.

W. Willett. "Vitamin A and Lung Cancer." *Nutrition Reviews*, Vol. 48, pgs 201–211, 1990.

R.G. Ziegler. "A Review of Epidemiologic Evidence that Carotenoids Reduce the Risk of Cancer." *Journal of Nutrition*, Vol. 119, pgs. 116–122, 1989.

The Alpha-Tocopherol, Beta-Carotene Cancer Prevention Study Group. "The Effect of Vitamin E and Beta Carotene on the Incidence of Lung Cancer and Other Cancers in Male Smokers." *The New England Journal of Medicine*,

Vol. 330, pgs. 1029–1035, 1994. (Note: This is interesting in that protective benefits of taking vitamin E supplements were not seen while beta carotene may have worsened the situation. The authors concluded that taking just supplements needs to be further evaluated and that possibly these vitamins are not the active cancer-inhibiting components of fruits and vegetables that are observed to protect in observational studies.)

D.L. Davis, H.P. Freeman. "Essay: An Ounce of Prevention." *Scientific American,* September, 1994, pg. 112.

T. Hirayama. "Diet and Cancer." *Nutrition and Cancer,* Vol. 1, pgs. 67–81, 1979.

CHAPTER 5

The biological activities of plants and natural products have been known since classic times. There are numerous modern and classic texts and review articles available on the medicinal and toxicological properties of plants and their by-products. An interesting observation is that rarely are the two combined. The readers should also be aware that many modern pharmaceutical drugs are derived from plant preparations which encompass the fields of pharmacognosy and natural products chemistry.

J. M. Kingsbury. *Poisonous Plants of the United States and Canada.* Englewood Cliffs, NJ: Prentice-Hall, 1964.

W.A.R. Thompson. *Herbs that Heal.* New York, Charles Scribner's Sons, 1976.

D.D. Buchman. *Herbal Medicine.* New York: Gramercy Publishing, N.Y., 1980.

G.E. Trease, W.C. Evans. *Pharmacognosy.* 12th Edition, London: Baillière-Tindall, 1988.

V.E. Tyler, L.R. Brady, J.E. Robbers. *Pharmacognosy.* 8th Edition, Philadelphia: Lea and Febiger, 1981.

B.J. Gurley, S.F. Gardner, L.M. White, P.L. Wang. "Ephedrine Pharmacokinetics after the Ingestion of Nutritional Supplements Containing *Ephedra sinica* (ma huang)." *Therapeutic Drug Monitoring,* Vol. 20, pgs. 439–445, 1998.

K.C. Huang. *The Pharmacology of Chinese Herbs.* Boca Raton, FL: CRC Press, 1993.

G.Q. Liu. "Chinese Natural Products and New Drugs." *Pharmaceutical News,* Vol. 2, pgs. 10–12, 1995.

L. Anderson, G.J. Higby. *The Spirit of Voluntarism. A Legacy of Commitment and Contribution. The United States Pharmacopeia 1820–1995.* Rockville, MD: United States Pharmacopeia, 1995. (Note: This provides an excellent annotated history of the development of pharmacy.)

R. Gregory, J. Flynn , P. Slovic. "Technological Stigma." *American Scientist,* Vol. 83, pgs. 220–223, 1995.

M.O. Amdur, J. Doull, C.D. Klaassen. *Casarett and Doull's Toxicology: The Basic Science of Poisons*, Fourth Edition. New York: Pergammon Press, 1991.

"An Interview with Bruce Ames." *OMNI*, pgs. 75–80, 103, 106, February, 1991.

B.N. Ames. "Six Common Errors Relating to Environmental Pollution." *Regulatory Toxicology and Pharmacology*, Vol. 7, pgs. 379–383, 1987.

B.N. Ames, L.S. Gold. "Too Many Rodent Carcinogens: Mitogenesis Increases Mutagenesis." *Science*, Vol. 249, 970–971, 1990.

B.N. Ames, R. Magaw, L.S. Gold. "Ranking Possible Carconogenic Hazards." *Science*, Vol. 271, pgs. 271–280, 1987.

B.N. Ames, M. Profet, L.S. Gold. "Dietary pesticides (99.99% all natural)." *Proceedings of the National Academy of Science USA*, Vol. 87, pgs. 7777–7781, 1990.

The following references support the general discussion of mycotoxin toxicology. I used these to illustrate foodborne poisons since I have worked with them in the past and have a firsthand familiarity with the literature. They are fully reviewed in the first text.

M.O. Amdur, J. Doull, C.D. Klaassen. *Casarett and Doull's Toxicology: The Basic Science of Poisons*, Fourth Edition. New York: Pergammon Press, 1991.

B. Ballantyne, T. Marrs, P. Turner. *General and Applied Toxicology.* New York: Stockton Press, 1993.

V.R. Beasley. *Trichothecene Mycotoxicosis: Pathophysiologic Effects, Volumes I and II.* Boca Raton, FL: CRC Press, 1989.

A.M. Bongiovanni. "An Epidemic of Premature Telarche in Puerto Rico." *Journal of Pediatrics*, Vol. 103, pgs. 245–246, 1983.

A.W. Hayes, ed. *Principles and Methods of Toxicology*, Third Edition. New York: Raven Press, 1994.

R.L. Maynard. "Toxicology of Chemical Warfare Agents." In A. W. Hayes, ed., *Principles and Methods of Toxicology*, Third Ed. New York: Raven Press, 1994, pgs. 1253–1286.

J.L. Shaeffer, J.K. Tyczkowska, J.E. Riviere, P.B. Hamilton. "Aflatoxin-impaired Ability to Accumulate Oxycarotenoid Pigments During Restoration in Young Chickens." *Poultry Science*, Vol. 67, pgs 619–625, 1988.

R.P. Sharma, D.K. Salunkhe. *Mycotoxins and Phytoalexins.* Boca Raton, FL: CRC Press, 1991.

S.F. Sundlof, J.E. Riviere, A.L. Craigmill. *Handbook of Comparative Pharmacokinetics and Residues of Pesticides and Environmental Contaminants in Animals.* Boca Raton, FL: CRC Press, 1995.

J.L. Casti. *Complexification: Explaining a Paradoxical World through the Science of Surprise.* New York: HarperCollins Books, 1994.

K.R. Foster, D.E. Bernstein, P.W. Huber, eds. *Phantom Risk. Scientific Inference and the Law.* Cambridge, MA: The MIT Press, 1993. (See legal discussions on toxic anxiety on pages 33 and 343 to 345. Note: I would strongly recommend this book for a more technical presentation of risk assessment that the authors say occur at the "interface of science and law.")

W. Spindell. "The Latent Neuropsychological Effects of Toxic Exposure." *Forensic Examiner,* Vol. 4, pg. 14, March/April 1995.

There are a number of references by Dr. Edward Calabrese introducing and arguing for the validity of the hormesis hypothesis. These include the *BELLE Newsletter* (Biological Effects of Low Level Exposure) published by Dr. Calabrese from the University of Massachusetts. Two good introductory references are these:

E. Calabrese, L.A. Baldwin. "Hormesis: U-Shaped Dose Responses and Their Centrality in Toxicology." *TRENDS in Pharmacological Sciences,* 22: 285–291, 2001.

E. Calabrese, L.A. Baldwin. "U-Shaped Dose-Responses in Biology, Toxicology, and Public Health." *Annual Review of Public Health,* 22: 15–33, 2001.

CHAPTER 6

The following references provide some interesting perspectives on the public's perceptions of risk and how chemical risk can be communicated:

American Veterinary Medical Association Food Safety Workshop. *Journal of the American Veterinary Medical Association.* Vol. 201, pgs. 227–266, 1992.

Office of Technology Assessment. *Pesticide Residues in Food: Technologies for Detection.* Washington, D.C., U.S. Government Printing Office, 1988.

National Research Council. *Meat and Poultry Inspection: The Scientific Basis of the Nation's Programs.* Washington, D.C., National Academy Press, 1985.

N. Neil, T. Malmfors, P. Slovic. "Intuitive Toxicology: Expert and Lay Judgements of Chemical Risks." *Toxicologic Pathology,* Vol. 22, pgs. 198–201, 1994.

P. Slovic. "Perception of Risk." *Science,* Vol. 236, pgs. 280–285, 1987.

P. Slovic, N.N. Kraus, H. Lappe, H. Letzel, T. Malmfors. "Risk Perception of Prescription Drugs: Report on a Survey in Sweden." *Pharmaceutical Medicine,* Vol. 4, pgs. 43–65, 1989.

Texas A&M Food Safety Workshop. *Journal of the American Veterinary Medical Association,* Vol. 199, pgs. 1714–1721, 1991.

CHAPTER 7

Proceedings of the International Symposium on Chemical Mixtures: Risk Assessment and Management. The Jerry F. Stara Memorial Symposium. *Toxicology and Environmental Health*, Vol. 5, #5, 1989.

S.L. Friess. "Commentary: The Use of Toxicological Research Information: How and by Whom." *Fundamental and Applied Toxicology*, Vol. 26, pgs. 151–155, 1995. (Note: This is an excellent commentary on the problems that scientists themselves and regulators have in using one another's data.)

A.D. Kligerman, R.E. Chapin, G.L. Erexson, D.R. Germolec, P. Kwanyuen, R.S.H. Yang. "Analyses of cytogenetic damage in rodents following exposure to simulated groundwater contaminated with pesticides." *Mutation Research*, Vol. 300, pgs. 125–134, 1993.

M. Lappé. *Evolutionary Medicine: Rethinking the Origins of Disease.* San Francisco: Sierra Club Books, 1994.

A. Lieberman. *Facts versus Fears: A Review of the 20 Greatest U.S. Health Scares of Recent Times.* American Council on Science and Health, April, 1997.

R.M. Nesse, G.C. Williams. *Why We Get Sick: The New Science of Darwinian Medicine.* New York: Times Books, 1994. (Note: The quote on longevity is from pgs. 136–137. This book is an excellent overview of this field and gave this author some interesting insights on how chemicals interact with the body in the context of evolution.)

J.H. Ross, J.H. Driver, R.C. Cochran, T. Thongsinthusak, R.I. Krieger. "Could Pesticide Toxicology Studies be More Relevant to Occupational Exposure?" *Annals of Occupational Hygiene*, 45: S5–S17, 2001.

C. Woodyard. "Fliers Fume over Planes Treated with Pesticides." *USA Today*, September 10, 2001.

R.S.H. Yang, ed. *Toxicology of Chemical Mixtures: Case Studies, Mechanisms and Novel Approaches.* New York: Academic Press, 1994. (Note: This book is an excellent source book on the science behind the study of chemical mixture toxicology.)

The following selections provide a broad coverage of MCS-type illnesses and related research:

Proceedings of the Association of Occupational and Environmental Clinics (AOEC) Workshop on Multiple Chemical Sensitivity. *Toxicology and Environmental Health*, Vol., 8, No. 4, July–August, 1992. (The definition for MCS is from the article by Ross on pg. 21. This book is the proceedings of a technical conference on MCS. This was a major reference source for technical material in this chapter. It should be contrasted with popular books whose message is very different.)

B. Arvidson. "A Review of Axonal Transport of Metals." *Toxicology*, 88: 1–14, 1994.

I.R. Bell, C.S. Miller, G.E. Schwartz. "An Olfactory Limbic Model of Multiple Chemical Sensitivity Syndrome. Possible Relationship to Kindling and Affective Disorders." *Biological Psychiatry*, Vol. 32, pgs. 218–242.

D.C. Dorman, M.F. Struve, B.A. Wong. "Pharmacokinetic Factors that Influence Manganese Delivery to the Brain." *CIIT Activities*, 21 (7/8): 1–8, July–August, 2001.

K. Hyams, F. Wignall, R. Roswell. "War Syndromes and Their Evaluation: from the U.S. Civil War Through the Persian Gulf War." *Annals Internal Medicine*, 125: 398–405, 1996.

D.J. Kutsogiannis, A.L. Davidoff. "A Multiple Center Study of Multiple Chemical Sensitivity Syndrome." *Archives of Environmental Health*, 56: 196–207, 2001.

L. Lawson. *Staying Well in a Toxic World. Understanding Environmental Illness, Multiple Chemcial Sensitivities, Chemical Injuries and Sick Building Syndrome.* Chicago: The Noble Press, 1993.

T.G. Randolph, R.W.Moss. *An Alternative Approach to Allergies.* New York: Lippincott and Crowell, 1980.

T.G. Randolph. *Human Ecology and Susceptibility to the Chemical Environment.* Springfield, IL: Thomas, 1962.

G.D. Ritchie, J.R. Still, W.K. Alexander, A.F. Nordholm, C.L. Wilson, J. Rossi III, D.R. Mattie. "A Review of the Neurotoxicity Risk of Selected Hydrocarbon Fuels." *Journal of Toxicology and Environmental Health, Part B.*, 4: 223–312, 2001.

L.M. Seltzer. "Building Related Illnesses." *Journal of Allergy and Clinical Immunology*, 94: 351–361, 1994.

F. Setterberg, L. Shavelson. *Toxic Nation: The Fight to Save Our Communities from Chemical Contamination.* New York: John Wiley & Sons, 1993.

E.E. Sikorski, H,M, Kipen, J.C. Selner, C.M. Miller, K.E. Rodgers. "The Question of Multiple Chemical Sensitivity." *Fundamental and Applied Toxicology*, Vol. 24, pgs. 22–28, 1995. (Note: This is a critical summary of a Society of Toxicology Rountable Discussion addressing the issue of MCS.)

H. Tjälve, J. Henriksson. "Uptake of Metals in the Brain via Olfactory Pathways." *NeuroToxicology*, 20: 181–195, 1999.

S. Wessely. "Ten Years On: What Do We Know About the Gulf War Syndrome?" *Clinical Medicine*, 1: 28–37, 2001.

The following references illustrate the nature of the so-called "Environmental Estrogen" debate. The first two discuss the observation that there is evidence that male sperm counts have been decreasing over the past fifty years.

E. Carlsen, A. Giwereman, N. Keiding, N.E. Skakkebaek. "Evidence of Decreasing Quality of Semen During the Past 50 Years." *British Medical Journal,* Vol. 305, pgs. 609–613, 1992.

R.M. Sharpe. "Declining Sperm Counts in Men—Is There an Endocrine Cause?" *Journal of Endocrinology,* Vol. 136, pgs. 357–360.

A. L. Lister, G.J. VanDer Kraak. "Endocrine Disruption: Why is it So Complicated?" *Water Quality Research Journal Canada,* 36: 175–190, 2001.

The next papers discuss another observation that various sexual abnormalities have been observed in wild populations of some toads, fish, and other reptiles. A link is then postulated that chemicals must be the cause of both. The argument is broadened when some pesticide classes are then thought to be partially responsible for breast cancer in women.

S. Begley, with D. Glick. "The Estrogen Complex." *Newsweek,* March 21, 1994.

C. Carey. "Hypothesis Concerning the Causes of the Disappearance of Boreal Toads from the Mountains of Colorado." *Conservation Biology,* Vol. 7, pgs. 355–362, 1993.

T. Colborn, C. Clement, eds. *Chemically-Induced Alterations in Sexual and Functional Development: The Wildlife/Human Connection.* Princeton, N.J.: Princeton Scientific Publishing Co., 1992.

R. Lewis. "PCB Dilemma." *The Scientist,* 15 (6): 1, 14–17, March 19, 2001.

D. Murray, J. Schwartz, S.R. Lichter. *It Ain't Necessarily So.* New York: Rowan and Littlefield, 2001.

J. Raloff. "The Gender Benders, Are Environmental 'Hormones' Emasculating Wildlife?" *Science News,* Vol. 145, pgs 24–27.

The final three references are important. The first is a major study, judged by many as the best yet conducted, which refutes the suggestion that pesticides such as DDT are at all related to the incidence of breast cancer in women today. The second is a balanced overview of the subject. The third is a candid critique of epidemiology by epidemiologists that stresses how weak correlations made between an observed trend and some chemical can be very dangerous and misleading. It would be good if more people heeded these warnings.

N. Krieger, M.S. Wolff, R.A. Hiatt, M. Rivera, J. Vogelman, N. Orentreich. "Breast Cancer and Serum Organochlorines: A Prospective Study Among White, Black, and Asian Women." *Journal of the National Cancer Institute,* Vol. 86, pgs. 589–599, 1994.

R. Stone. "Environmental Estrogens Stir Debate." *Science,* Vol. 265, pgs. 308–310, 1994.

G. Taubes. "Epidemiology Faces Its Limits." *Science,* Vol. 269, pgs. 164–169, 1995.

CHAPTER 8

American Veterinary Medical Association and National Milk Producers Federation. *Milk and Dairy Beef Residue Prevention: A Quality Assurance Protocol.* July, 1991.

National Academy of Sciences. *The Effects on Human Health of Subtherapeutic Use of Antimicrobials in Animal Feeds.* Washington, D.C., 1980.

National Academy of Sciences. *Human Health Risks with the Subtherapeutic Use of Penicillin or Tetracycline in Animal Feed.* Washington, D.C., 1989.

National Research Council. *The Use of Drugs in Food Animals: Benefits and Risks.* Washington, D.C., 1999.

Office of Management and Budget. *Use of Bovine Somatotropin (BST) in the United States: Its Potential Effects, A Study Conducted by the Executive Branch of the Federal Government.* January, 1994.

J.C. Juskevich, C.G. Guyer. "Bovine Growth Hormone: Human Food Safety Evaluation." *Science,* Vol. 249, pgs. 875–884, August 24, 1990.

J.E. Riviere, A.L. Craigmill, S.F. Sundlof. *Comparative Pharmacokinetics and Residues of Veterinary Antimicrobials.* Boca Raton, FL: CRC Press, 1991.

J.E. Riviere, A.L. Craigmill, S.F. Sundlof: "The Food Animal Residue Avoidance Databank (FARAD): An Automated Pharmacologic Databank for Drug and Chemical Residue Avoidance." *Journal of Food Protection,* Vol 49, pgs. 826–830, 1986.

K.L. Ropgs. "No Human Risks: New Animal Drug Increases Milk Production." *FDA Consumer,* pgs. 24–27, May, 1994.

P. Willeberg. "An International Perspective on Bovine Somatotropin and Clinical Mastitis." *Journal of the American Veterinary Medical Association,* Vol. 205, pgs. 538–541, August 15, 1994.

W.W. Williams, D.B. Weiner. *Biologically Active Peptides: Design, Synthesis, and Utilization.* Lancaster, PA: Technomic Publishing Co., 1993.

CHAPTER 9

For a basic review of heredity, evolution, and natural selection, the previously quoted texts on "Darwinian Medicine" by either Nesse and Williams or Lappé could be consulted, as they both introduce these concepts. For those readers with minimal background in genetics and molecular biology, which is the foundation of biotechnology, a recent encyclopedia or introductory textbook of biology is the best first step. Alternatively, some of the following texts are appropriate.

R. Dawkins. *The Selfish Gene.* Oxford: Oxford University Press, 1976.

H. Gershowitz, D.L. Rucknagel, R.E. Tashian, eds. *Evolutionary Perspectives and the New Genetics.* New York: Plenum Press, 1985.

R.J. MacIntyre. *Molecular Evolutionary Genetics*. New York: Plenum Press, 1985.

E. Mayer. *The Growth of Biological Thought: Diversity, Evolution and Inheritance*. Cambridge: The Belknap Press of Harvard University Press, 1982. (Note: This is a classic and influential overview of the philosophy underpinning biology which includes an excellent and pertinent introduction to genetics and heredity.)

N.V. Rothwell. *Understanding Genetics*, 4th Ed. New York: Oxford University Press, 1988.

J.K. Setlow, ed. *Genetic Engineering*. Vols. 1–15. New York: Plenum Press, 1979–1993.

J.D. Watson. *Molecular Biology of the Gene*. New York: W.A.Benjamin, 1965.

The sequence of the human genome was published by a private team of scientists working for Celera Genomics in:

Venter, et al. "The Sequence of the Human Genome." *Science*, 291: 1304–1351, February 16, 2001

and by the publicly supported Human Genome Project in:

Ramser et al., "A Physical Map of the Human Genome." *Nature*, 409: 934–941, February 15, 2001.

For an update on some of the latest developments in plant biotechnology:

"Emerging Plant Science: Frontiers in Biotechnology." *Science*, Vol. 268, pgs. 653–719, 1995.

P.H. Abelson, P.J. Hines. "The Plant Revolution." *Science*, 285: 367–389, 1999.

J. Finnegan, D. McElroy. "Transgene Inactivation: Plants Fight Back!" *Bio/Technology*, Vol. 12, pgs. 883–887, 1994.

P.J. Hines, J. Marx. "The Endless Race Between Plant and Pathogen." *Science*, 292: 2269–2276, 2001.

T.J. Hoban , P.A. Kendall: "Project Summary: Consumer Attitudes about Food Biotechnology." North Carolina State University, Raleigh, NC.; U.S.D.A. Extension Service Project 91-EXCA-3-0155, 1993.

E. Hodgson. "Genetically Modified Plants and Human Health Risks: Can Additional Research Reduce Uncertainties and Increase Public Confidence?" *Toxicological Sciences*, 63: 153–156, 2001.

R. Hoyle. "EPA Okays First Pesticidal Transgenic Plants." *Bio/Technology*, Vol. 13, pgs 434–435, 1995.

S. Kilman. "Use of Genetically Modified Seeds by U.S. Farmers Increases 18%." *Wall Street Journal*, July 2, 2001

H.A. Kuiper, G.A.Kleter, H.P.J.M. Noteborn, E. Kok. "Assessment of the Food Safety Issues Related to Genetically Modified Foods." *The Plant Journal*, 27: 503–528, 2001.

C.C. Mann. "Biotech Goes Wild." *MIT Technology Review*, 35–43, July/August, 1999.

M. Marvier. "Ecology of Transgenic Crops." *American Scientist*, 89: 160–167, 2001.

T.G. Terre, J.D. Green. "Issues with Biotechnology Products in Toxicologic Pathology." *Toxicologic Pathology*, Vol. 22, pgs. 187–193, 1994.

L.L. Wolfenbarger, P.R. Phifer. "The Ecological Risks and Benefits of Genetically Engineered Plants." *Science*, 290: 2088–2093, 2000.

B. Zechendork. "What the Public Thinks about Biotechnology." *Bio/Technology*,Vol. 12, pgs. 875, 1994.

For a complete analysis of the EPA's actions over the StarLink® Biotech Bt Corn scare of 2000, see the EPA's regulation which reauthorized this product after scientific review:

"Biopesticides registration action document. *Bacillus thuringiensis* plant-incorporated protectants." Oct 16, 2001.

M. Crichton. *Jurassic Park*. New York: Ballantine Books, 1990. (Note: This is an excellent example of how a well-written science fiction novel, using what looks like a plausible extension of genetic engineering, can weave a tale of horror and the possibility of a dark future emanating from the laboratory of mad scientists. It, and the three movies it spawned, also highlights the perceived arrogance of scientists who think that they truly understand the operation of the world. It is a great novel, but it is not a textbook of biotechnology.)

D. Waltner-Toews. *Food, Sex and Salmonella: The Risks of Environmental Intimacy*. Toronto: New Canada Press Limited, 1992. (Note: This is a very entertaining look at the problems which really are associated with eating food. It actually is the "flip-side" of the present book since it assumes that there are no real risks from pesticide residues and instead focuses on diseases that are killing people today.)

CHAPTER 10

"Botulinum Toxin Type A." United States Pharmacopeial Convention: Drug Information for the Health Care Professional. Vol. I, 22nd edition, 650–655, 2002.

C.F. Chyba. "Biological Security in a Changed World." *Science*, 293: 2349, 2001.

R.L. Maynard. "Toxicology of Chemical Warfare Agents," *Principles and Methods of Toxicology, 3rd Ed.* (ed. A.W. Hayes), New York: Raven Press, pgs. 1253–1286, 1994.

R. Robinson. "Food Safety and Security: Fundamental Changes Needed to Ensure Safe Food." *United States General Acounting Office.* Washington, D.C., October 10, 2001, GAO-02-47T.

A. Rogers, M. Isikoff, D. Klaidman, D. Rosenberg, E. Check, F. Guterl, E. Conant, J. Contreras, S. Downey. "Unmasking Bioterror." *Newsweek,* pgs. 22–29, October 8, 2001.

L. Szinicz, S.I. Baskin. "Chemical and Biological Agents," Chapter 35, *Toxicology,* San Diego: Academic Press, pgs. 851–877, 1999.

A.P. Watson, T.D. Jones, J.D. Adams. "Relative Potency Estimates of Acceptable Residues and Reentry Intervals After Nerve Agent Release." *Ecotoxicology and Environmental Safety,* 23: 328–342, 1992.

CHAPTER 11

C. Anderson. "Cholera Epidemic Traced to Risk Miscalculation." *Nature,* Vol. 354, pg. 255, 1991. (Note: Source for data on the Peruvian cholera epidemic which resulted when chlorine was removed from the water supply.)

R.D. Fleischmann, M.D. Adams, O. White, R.A. Clayton, E.F. Kirkness, A.R. Kerlavage, C.J. Bult, J.-F. Tomb, B.A. Dougherty, J.M. Merrick, K. McKenny, G. Sutton, W. FitzHugh, C. Fields, J.D. Gocayne, J. Scott, R. Shirley, L.I.Liu, A. Glodek, J.M. Kelley, J.F. Weidman, C.A. Phillips, T. Spriggs, E. Hedblom, M.D. Cotton, T.R. Utterback, M.C. Hanna, D.T. Nguyen, D.M. Saudek, R.C. Brandon, L.D. Fine, J.L. Fritchman, J.L. Furhman, N.S.M. Geoghagen, C.L. Gnehm, L.A. McDonald, K.V. Small, C.M. Fraser, H.O. Smith, J.C. Venter. "Whole-Genome Random Sequencing and Assembly of *Haemophilus influenza* Rd." *Science,* Vol. 269, pgs. 496–512, 1995. (Note: This is the paper in which the entire genetic code, all 1,830,137 nucleotides, that defines this bacteria is published. This is the first report of a complete genome sequence for a free-living organism and signifies a state of maturation for the field of molecular biology and genetic engineering.)

This final chapter is based upon all that precedes it. Pivotal sources have been cited in these earlier chapters. However, there are a number of additional general books and so-called "classics" on science and its limits and future challenges that have shaped my thoughts. These are listed here in alphabetical order:

Anonymous. "Bovine Spongiform Encephalopathy," Fact Sheets. *USDA, A.P.H.I.S. Veterinary Services,* January, 2001; and *C.D.C.* January, 2001.

Anonymous. "Food Safety Symposium—Post Harvest." *Journal of the American Veterinary Medical Association*, 213: 1729–1751, 1998.

C.F. Amábile-Cuevas, M. Cárdenas-Gariá, M. Ludgar. "Antibiotic Resistance." *American Scientist*, Vol. 83, pgs. 320–329, 1995.

E. Bingham, D.P. Rall. "Preventive Strategies for Living in a Chemical World." *Annals of the New York Academy of Sciences*, Vol. 837, 1997.

J.C. Burnham. *How Superstition Won and Science Lost*. New Brunswick, NJ.: Rutgers University Press, 1987.

R.N. Butler, A.G. Bearn. *The Aging Process: Therapeutic Implications*. New York: Raven Press, 1985.

J.L. Casti. *Paradigms Lost: Images of Man in the Mirror of Science*. New York: William Morrow, 1989.

J.L.Casti. *Searching for Certainty: What Scientists Can Know About the Future*. New York: William Morrow, 1990.

L. Garrett. *The Coming Plague: Newly Emerging Diseases in a World Out of Balance*. New York: Farrar, Straus and Giroux, 1994.

B.J. Gerloff. "Are We Sure We Are Doing This Right" *Journal of the American Veterinary Medical Association*. Vol. 207, pgs 154–155, 1995.

D.C.Grossman, H. Valtin. "Great Issues for Medicine in the 21st Century." *Annals of the New York Academy of Sciences*, Vol. 882, 1999.

R. Harré. *The Principles of Scientific Thinking*. Chicago: The University of Chicago Press, 1970.

G. Holton. *Thematic Origins of Scientific Thought: Kepler to Einstein*, Revised Edition. Cambridge, MA: Harvard University Press, 1988.

G. Holton. *Science and Anti-Science*. Cambridge, MA: Harvard University Press, 1993.

C. Howson, P. Urbach. *Scientific Reasoning: The Baysean Approach*. LaSalle, IL: Open Court, 1989.

P.W. Huber. *Galileo's Revenge: Junk Science in the Courtroom*. New York: Basic Books, 1991.

T.S. Kuhn. *The Structure of Scientific Revolutions*, Second Edition, Enlarged. Chicago: University of Chicago Press, 1970.

R. Lewis. "Portals for Prions?" *The Scientist*, 15 (15): 1, 21,23; July 23, 2001.

M. LaFollette, ed. *Quality in Science*. Cambridge, MA: The MIT Press, 1982.

L.A. Muir Gray. "Postmodern Medicine." *Lancet* 354: 1550–1553, 1999.

M.S. Nawaz, B.D. Erickson, A.A. Khan, S.A. Khan, J.V. Pothuluri, F. Rafii, J.B. Sutherland, R.D. Wagner, C.E. Cerniglia. "Human Health Impact and Regulatory Issues Involving Antimicrobial Resistance in the Food Production Environment." *Regulatory Research Perspectives*, 1: 1–10, 2001.

S.R. Palumbi. "Humans as the World's Greatest Evolutionary Force." *Science*, 293: 1786–1790, 2001.

R. Preston. *The Hot Zone*. New York: Random House, 1994.

J. Redfield. *The Celestine Prophesy; An Adventure.* New York: Warner Books, 1993.

L.L. Smith. "Key Challenges for Toxicologists in the 21st Century." *Trends in Pharmacological Sciences,* 22: 281–285, 2001.

R. Stone. "A Molecular Approach to Cancer Risk Assessment." *Science,* Vol. 268, pgs. 356–357, 1995.

L. Tollefson. "Chemical Food Borne Hazards and Their Control." *The Veterinary Clinics of North America—Food Animal Practice,* 15:1, 1999.

E.O. Wilson. *Sociobiology: The New Synthesis.* Cambridge, MA: Harvard University Press, 1975.

E.O. Wilson. *On Human Nature.* Cambridge, MA: Harvard University Press, 1978.

J. Zinsstag, M. Weiss. "Livestock Diseases and Human Health." *Science,* 294: 477, 2001.

Appendix A

There are numerous technical references available to review how risk assessment is conducted and how tolerances are set by the federal government. A few of these are listed below. Numerous laws and legal tomes have likewise been written. The establishment of such tolerances is also a constantly changing process, tweeked by both science and law. The purpose of this section is to illustrate the concept, not present the regulations.

R. Adams, ed. *Veterinary Pharmacology and Therapeutics.* Seventh Edition, Ames, IA: Iowa State University Press, 1995.

B. Ballantyne, T. Marrs, P. Turner. *General and Applied Toxicology.* New York: Stockton Press, 1993.

A.W. Hayes, ed. *Principles and Methods of Toxicology,* Third Edition. New York: Raven Press, 1994.

National Research Council. *Improving Risk Communication.* Washington, D.C.: National Academy Press, 1989.

National Research Council. *Issues in Risk Assessment.* Washington, D.C.: National Academy Press, 1993.

National Research Council. *Pesticides in the Diets of Infants and Children.* Washington, D.C.: National Academy Press, 1993. (Note: See the Executive Summary on pgs 1–12 for a concise discussion on the adequacy of the existing uncertainty factors).

D. Waltner-Toews, S.A. McEwen. "Human Health Risks From Chemical Contaminants in Foods of Animal Origin." *Preventative Veterinary Medicine* Vol. 20, pgs. 158–247, 1994.

The selected toxicology review in Appendix A is a distillation of material in a technical book on pesticide residues which I recently published. Since this reviewed, referenced, and edited material was completed in 1995, it provides a timely overview of pesticide toxicology. I thank my coauthors, Drs. Craigmill and Sundloff, for their original input into this material. The reader is directed to this source for a more in-depth treatment and full bibliographical citations:

S.F. Sundlof, J.E. Riviere, A.L. Craigmill. *Handbook of Comparative Pharmacokinetics and Residues of Pesticides and Environmental Contaminants in Animals*. Boca Raton, FL: CRC Press, 1995.

In addition, there are numerous toxicology textbooks that address these chemicals in great detail. The following is a list of some selective readings:

C.D. Klaassen. *Casarett and Doull's Toxicology: The Basic Science of Poisons*, Sixth Edition. New York: McGraw Hill, 2001.

M.O. Amdur, J. Doull, C.D. Klaassen. *Casarett and Doull's Toxicology: The Basic Science of Poisons*, Fourth Edition. New York: Pergammon Press, 1991.

B. Ballantyne, T. Marrs, P. Turner. *General and Applied Toxicology*. New York: Stockton Press, 1993.

S. Budavari. *The Merck Index*, Eleventh Edition, Rahway: Merck and Co., 1989.

J.E. Chambers, P.E. Levi, eds. *Organophosphates: Chemistry, Fate and Effects*. New York: Academic Press, 1992.

A.W. Hayes, ed. *Principles and Methods of Toxicology*, Third Edition. New York: Raven Press, 1994.

R.C. Honeycutt, E.W. Day, Jr., eds. *Worker Exposure to Agrochemicals*, Boca Raton: Lewis Publishers, 2001.

J.K. Marquis. *Contemporary Issues in Pesticide Toxicology and Pharmacology*. Basel: Karger, 1986.

G.W.A. Milne. *CRC Handbook of Pesticides*. Boca Raton, FL: CRC Press, 1995.

Index

A

absorption
  BST, 115
  fat soluble chemicals and, 29
  IGF-I, 115
  kidneys and, 26–27
  liver and, 26–27
  peptides, 113–114
  tissue, movement to, 27
absorption vs exposure, 26
acetylcholinesterase enzyme
  inihibition of, 172
  organophosphates and, 97
active transport, 27
additive interactions, 95
  mixtures, 98
additive toxic effects of chemicals, 94
adverse reaction, risk assessment and, 12
aflatoxins, 71, 76
  food contamination and, 139
  food delivery, 142
  peanuts and, 74
aged persons. *See* older persons
Agent Orange, 137, 138, 175
agents
  biochemical/chemical, relevant to bioter-
    rorism and warfare, 136
  definitions, 37
aging, senescence, 98–99

airline use of pesticides, 95
Alar, 49, 72, 92
alcohol
  analgesics, liver toxicity and, 96
  cancer rates and, 60
  cirrhosis of the liver, 59
  consumption, smoking and, 96
  dioxin comparison, 28
  dose-response and, 27
  interactions and, 30
  liver and, 28–29
aldrin, 49, 166
  phase out, 42
alkaloids, hemlock, 70
allergic reactions, 32
  peanuts, 74
allergies
  StarLink corn, 130
  transgenic plants, 130
alpha-carotene, 65
American agricultural dominance, 41
Ames test for chemical mutagenicity, 73
amino acids
  kidney excretion, 27
  peptides, 113
amnioglycoside drugs, animal testing and,
    14
amygdalin, 72
analgesics, liver and, 96

analytical chemistry
 advances in, 33
 DES Proviso and, 34
Anderson, Dr. Phillip W., 35
animal testing. *See also* laboratory rats
 anmioglycoside drugs, 15
 dose and, 26
 drug sensitivity and, 14
 exaggerated responses, 14
 NIH approval, 15
 rats inbred, 14
 selection of animals, 14
 in vitro assays, 26
antagonistic interactions, 95
anthrax, 139
 vaccine risks, 12
antibiotics, 37
 BST-treated cows, 116
 drug-resistant bacteria potential, 118
 feed conversion and, 118
 kanamycin, resistance in plants, 131
 milk
  residue, 117
  testing policies, 117
 tetracycline, 98
anticholinesterase insecticides, 171
antifungal drugs, 37
antimicrobial drugs, 37
 bacterial resistance, 145–146
antioxidants, receptors and, 30–31
antivirals, 37
apples
 synthesizing from scratch, 72
 warning lable, 73
aquatic animals, polluted waters and, 107
argument depth of studies, 99
Aspergillus flavus, 76
Aspergillus parasiticus, 76
atrazine, 19
 farmers and, 62
audited protocols, toxicology tolerances, 161
Avery, Dennis
 cancer deaths and, 61
 DDT withdrawal and, 42

B
background disease, 15
 birth defects, 17
 linking to drug or chemical exposure, 16

bacterial contamination vs drug residues, 34–35
bacterial food poisoning, 131–132
bacterial mutagencity, in vitro studies, 31
Balkan nephropathy, ochratoxin and, 76
Belgian dioxin crisis, 108
bell-shaped distribution, statistics, 5
benchmark dose, 162
Bendectin antinausea drug, 16–17
beta-carotene, 65
 blood concentrations, 65
beta-cyrptoxanthin, 65
Bhagwan Shree Rajneesh cult,
 food poisoning, 139
bimodal population, statistics and, 5
Bin Laden, Osama, chemical warfare and,
 139
bioassays, cancer, 160
bioavailability
 bioavailable dose, 26
 in vitro studies and, 31
bioengineered food, super weeds, 131
biological activity, new insecticides, 92
biological pathogens, biosecurity and, 135
biological warfare
 history of, 137
 saliva from rabid dogs, Da Vinci, 137
biosecurity, 137. *See also* chemical defense
biotechnology
 advances in, 32
 developments in, 123
 plant biotechnology, 123
 plant hybrids, insect resistance and, 52
bioterrorism, 135
 anthrax, 139
 botulinum toxin, 141
 evidence of, 140
 inspections, 143
 milk in tanker trucks, 143
biotransformation of chlorinated hydrocarbon insecticides, 168
birds, DDT and, 39
birth defects
 Bendectin antinausea drug, 17
 percentage of occurence, 17
 vitamin A example, 23
bleopharospasms, Botox and, 141
Boer War, Rheumatic Condition, 100
Botox, 141

botulinum toxin
  food bioterrorism and, 141
  food contamination and, 139
  food delivery, 142
  natural outbreak of botulism of 1998
    and 1999, 142–143
  Red Army Faction, 141
bovine growth hormone, 112. *See also* BST
bovine insulin, 114
breast cancer, 107
  DDT and, 107–108
  exposure and, 62
  PCBs and, 107–108
broccoli consumption, cancer and, 66
Brockovich, Erin, 93
BSE (Bovine Spongiform Encephalitis), 146
BST (bovine somato-tropin), 112
  antibiotics and, 116
  approving regulatory agencies, 112
  codon sequence, DNA, 124
  economic issues, 115–116
  Escherichia coli and production, 127
  human interaction and, 115
  injection, 114
  insulin comparison, 113
  mastitis in cows, 116
  in milk, 74
  natural, as human growth hormone, 114
  pasteurization and, 115
  presence in milk, 113
Bt (Bacillus thuringiensis), 126
Bt corn (transgenic), 127
bubonic plague, biological warfare, 137

C
cabbage, chemical makup, 72
caffeic acid, 72, 74
Calabrase, Dr. Edward, low-dose exposure
    benefits, 66–67
Calgene FlavrSavr tomato, 127
cancer
  alcohol and, 60
  bioassay, 160
  breast, 107
  causes, Doll and Peto, 61–62
  colon cancer, fruits & vegetables and, 65
  death rate and, 59
    aged persons and, 60
    Census Bureau, 59

  lung cancer, 60
  digestive system, 60
  esophageal, mycotoxins and, 76
  flavenoids and, 64
  fruits & vegetables
    broccoli, 66
    consumption and prevention, 64
  high-dose treatment and, 32
  indole-3-carbinol and, 64
  lung, 60
  older persons, 78
  reproductive system, 60
  skin cancer, 60
  sulforaphane and, 64
  survival rates, 61
  testicular, increase in, 107
  tobacco and, 29–30
  vegetarians, 64–65
  vitamin A example, 23
cancer-causing chemicals
  oncogenes, 16
  sensitivity in inbred rats, 16
carbamates, 38, 170–174
carbaryl, 49, 171
carbofuran, 49, 171
carcinogenic cells, immune response to
    selves, 31
carcinogens
  animal, food supply and, 33–34
  drug residue and, DES Proviso, 34
  natural, 28
  peanut mold, 74
  zero tolerance policy, Delany Clause, 34
carotenoids, 65
  fruits & vegetables and, 65
Carson, Rachel, 38–40
catechol, 74
causal inference, 4
causation and relationships, 63
CDC (Centers for Disease Control), 135
celery, natural toxins, 74
cell culture studies, in vitro studies, 31
cellular phones, 92
Census Bureau
  death statistics, 57–58
  life-expectancy, 57–58
  vital statistics, 57
chaconine, 52
chance, statistics and, 5, 9

chemical defense, 137. *See also* biosecurity
chemical mixtures, 94
  additive effects, 98
  interactions and, 96
  synergistic effects, 98
  toxicity, 94
  toxicology of, 99
chemical mutagenicity, Ames test for, 73
chemical sensitivity, 101–102
chemical warfare
  Agent Orange, 137
  bin Laden, Osama, 139
  foodborne attack, 142
  harmful concentrations, testing, 143
  history of, 137
  Iraq, 138
  mustard gas shells, 138
  mycotoxins, 142
  nerve agents, gases, 141
  sarin, 171
  soman, 171
  tabun, 138, 171
  VX, 142
  WWI, 137
  WWII, 138
chemicals
  chlorinated, 108
  detection ability, increase, 32
  exposure, absorption and, 26
  fat soluble, 29
  fears' reversal of progress, 70
  interactions, 29
  mechanism of action, 32
  natural food, presence in, 69
  natural protection from, 30
  need for, 25
  nomenclature phobia, 70
  persistent, 39
  pest treatment, 38
  plant survival and, 70
  presence of one contaminating others, 97
chemophobia, xiii
chloramben, 49
chlordane, 49, 166
  phase out, 42
chlorinated chemicals, 108
chlorinated hydrocarbons, 38, 39, 166–170
  biotransformation, 168

chemical classes, 168
  chlorine in drinking water and, 43
    waterbourne bacterial epidemic, 43
chlorinated organic chemicals, 43
chlorine, chemical warfare, 138
chlorophenoxy compounds, 176
cholera
  food contamination and, 139
  Peru epidemic of 1980s, 152
chromosomes, plant breeding and, 124, 125
chronic-wasting disease, deer and elk and, 146
Civil War, Soldier's Heart, 100
clinical ecology, 101
Clostridia botulinum, 141
  food poisoning and, 131–132
clusters (statistics), gentamicin testing in rats, 15
codon alphabet, DNA, 125
codons, DNA, 124
coffee, natural carcinogens, 78
coin flip example, statistics, 9
colon cancer
  cabbage, 66
  fruits & vegetables, 65
combat stress, illness and, 100
compulsive behavior, MCS and, 103
confidence interval, statistics, 7
coniine, hemlock and, 70
consumption
  comparing to exposure levels, 46
  fruits & vegetables
    cancer prevention and, 64
    rate of consumption, 63
  thresholds, assumptions for, 48
copy machine removal, xv
corn
  aflatoxins, 76
  borer elimination, 126
  Bt corn (transgenic), 127
  Cry9C enzyme, 126
  F. roseum, 77
  fusarium moniliforme contamination, 76
  zearalenone, 77
cotton crops, 49
cottonseed, aflatoxins, 76
cows
  BST, 112
  increased milk production, 112

cranberry scare of 1959, 92
crotoxyphos, 171
Cry9C enzyme, corn, 126
Cryptosporidium, Milwakee, 152
cyanide
    compounds in fruits, 72
    potency of, 71
cyanide grouping (CN), 71
cyanogenic glycosides, amygdalin, 72
cyclamates, 92
cyclodienes, 169
cyclosarin, Iraq, 138
cytogenic effects, drinking water, 96

D
daily exposure vs foodborne, 94–95
dairy products
    BST, 112
    tetracycline and, 98
Damned Lies and Statistics, xiii
Darwinian medicine, 98
data anlysis, death rate information, 59
DBCP, 49
DDT, 29, 39, 166–170, 168–169
    Avery, Dennis and, 42
    breast cancer and, 107–108
    cost of removing from market, 41
    DDE and, 168
    estrogen resemblance, 107
    fear fueling, 43–44
    Leishmaniasi's, 42
    malaria, mosquito-carried, 42
    Müller, Dr. Paul, 168
    phase out, 42
    Ray, Dixie Lee and, 41
    Society of Toxicology, 43
    World War II use, 41
death
    acute, organophosphates, 42
    cause, Census Bureau, 58
    death rate, 55
        modern medicine and, 61
        reporting, population and, 59
    infant mortality rates, 58
    natural causes *versus* drugs or chemicals, 4
    questions to pose concerning
        pesticides, 56
    statistics, Census Bureau, 57–58
        heart disease and, 58–59

deer, chronic-wasting disease, 146
Delaney Clause, 33–34
    DES Proviso and, 34
    mycotoxins and, 75–76
    zero tolerance policy, 34
Department of Defense, Gulf War
        Syndrome and, 100
depression, MCS and, 103
dermal exposure, organophosphates, 42
DES (diethlystilbesterol), high-dose vs low-
        dose studies, 30
    beef and, 92
detectable residue, EPA tolerances, 45
detection, toxicity and, 34
dibenzofurans, 166
dichlorvos, 171
dieldrin, 166
    phase out, 42
diet pills, side effects, 71
diispropylfluorophosphonate, 137
dinoseb, 49
dioxin, 166
    Belgian crisis, 108
    estrogen resemblance and, 107
    phase out, 42
    tolerance, alcohol comparison, 28
disease
    background, 15–16
    control over, Western society's ideas, 4
    spontaneous, 15–16
diseased animals, biological warfare, 137
DNA (deoxyribonucleic acid)
    chemically damaged, repairing, 31
    computer comparison, 126
    condons, 124
    genomics, 125
    overview, 124
DNA gene "chips," 31
Doll and Peto, *The Causes of Cancer*, 61–62
dose
    animal testing and, 26
    bioavailable dose, 26
    lower/safer assumption, 25
    Paracelsus and, 19
    RfD, 162
    suicide from OD of simple drugs, 24
    vitamin A and, 21
dose-response
    alcohol, 27

dose-response (*continued*)
    farm chemicals and, 25
    receptor binding and, 30
    toxicology and, 24
    validation, 26
dose-response curve, 23–24
dose-response models, benchmark dose, 162
dose-response paradigm
    allergic reactions, 32
    hypersensitivity, 32
drinking water
    chlorine in, 43
    cholera, 152
    ground water contamination, 96–97
    interactions, 96
drug allergies, 32
drug approval process
    genetic toxicity testing, 26
    immunotoxicity, 26
    relaxed requirements, 13
    removal from market, 13
    separating effects from disease, 13
    side effects, chance effects, 13
drug doses, phamacokinetics and, 15
drug sensitivity, 14. *See also* sensitivity
drug trials statistics, 9–14
    placebo effect, 10
drugs, injected, 26
DSMA, 49
Dutch Elm disease, DDT and, 39

E
economics, pesticide reduction and, 52
edible tissues, drug levels and, 34
elixirs of death, 39
elk, chronic-wasting disease, 146
encephalitis, viral, 42
endogenous compunds, receptor binding
        and, 30
endrin, phase out, 42
environmental chemicals, environmental
        estrogen and, 30
environmental epidemiology, 18
environmental estrogens, 106
    environmental chemicals and, 30
    estradiol and, 30
Environmental Illness, 99
EPA (Environmental Protection Agency)
    establishment of, 38

residue tolerances, 45
    tolerance setting, 164
ephedra alkaloids, side effects, 71
epidemiology
    defined, 56
    environmental, 18
Erin Brockovich, 93
Eschericia coli
    drug production and, 127
    food poisoning and, 131–132
esophageal cancer, mycotoxins and, 76
estradiol, environment estrogens and, 30
estragole, 74
estrogen
    DDT and, 107
    dioxin and, 107
    environmental, 30, 106
    male levels of, 108
    mycotoxins and, 76–77
    natural (*See also* estradiol)
        binding to, 30
    PCBs and, 107
European corn borer, 126
evolution of human body, 98
exposure
    absorption and, 26
    access and, 87
    breast cancer, 62
    comparing with actual consumption, 46
    daily vs food consumption, 94–95
    detection, pesticides, 103
    fat soluble chemicals and, 29
    hazards and risk, 87
    multiple exposure routes, pets, 95
    perceived, MCS and, 104
    psychology of, legal issues, 80
    RfD and, 163
    thresholds, paradigm shift, 88
    toxicology tolerance, 160
extrapolation
    mathematical models for, 83
    over-extrapolation, 99
    pitfalls, 22
    title of research and, 99

F
F. roseum, corn and, 77
false impressions of toxicity, 105
false positives, 9

FARAD (Food Animal Residue Avoidance Databank), 33
farmers
    animals, zearalenone in pigs, 77
    atrazine and, 62
    occupational exposure, 61–62
    residue safety, 33
fat soluble chemicals, 29, 167
FDA (Food and Drug Administration)
    BST approval and, 112
    guidelines, 12
    legal issues, 12
    length of approval process, 13
    milk testing spot checks, 117
    post-approval monitoring, 13
    produce classifications, 44
    random samplings, 44
    Residue Monitoring Program, 44
    residue regulation, 34
    tolerance setting, 164
    unregistered pesticides, 46
feed contamination, PCB in Michigan dairy
        cattle, 142
fertilizer, 52–53
Fischer rats, 14
flavenoids, cancer and, 64
FMD (Foot and Mouth Disease), 141, 147
folic acid, 65
Food, Drug and Cosmetic Act of 1938,
        33–34
food allergies
    MCS and, 102
    peanuts, 74
food science biases, 3
foodborne illness
    bacterial food poisoning, 131–132
    Bhagwan Shree Rajneesh cult, 139
formaldehyde off-gasing,
        building materials, 103
fruits & vegetables, 63. See also produce
    broccoli consumption, 66
    cancer
        prevention, 64
        vegetarians' cancer rate, 64–65
    carotenoids and, 65
    cyanide compounds in fruits, 72
fungi
    aflatoxins, 76
    ochratoxins and, 76

fungicides, 38
    food additive classification, 75
    peanut mold, 74–75
furans, phase out, 42
fusarium moniliforme, contamination of
        corn, 76

G
genetic toxicity testing, 26
genomics, 125
gentamicin, kidney damage and, 15
GLPs (good laboratory practices), 161
glycophosphate (Roundup), 51
golden rice, 126
grains
    aflatoxins, 76
    mycotoxin production, 75
grams per acre, pesticides, 51
grasses, mycotoxin production, 75
groundwater contamination, 96–97
Gulf War Syndrome, 100

H
HACCP (Hazard Analysis and Critical
        Control Points) inspection process,
        144
hair dyes, 92
half-life (T1/2), 165–166
halogenated hydrocarbons, 166
    metabolic conversion, 167–168
hazards
    exposure and risk, 87
    RfD and, 163
    toxicology and, 160
HCB (hexachlorobenzene), 168, 170
health, overall
    age of pesticides, 55
    status as of 2001, 56–63
heart disease, death rate and, 58–59
hemlock, Socrates and, 70
hepatitis
    Silent Spring and, 59
    viral, rates, 59
heptachlor, 166
    phase out, 42
herbal medicine, 69
herbicides, 38
    glycophosphate (Roundup), 51
    toxicology, 175–178

hexachlorobenzene, phase out, 42
hierarchical structure of reality, 35
high-dose studies
    environmental estrogen, 30
    mathematical models for extrapolation,
        83
    thresholds and, 48
homeostatic compensatory mechanisms
    sub toxic chemical exposure and, 22
hormesis, dose-response and, 24
    Calabrase, Dr. Edward and, 66–67
    Darwinian medicine and, 98
hormones, natural, 74
Hudson River cleanup, 107
human growth hormone, 114
human interpretation, risk assessment and,
    11
human population, wildlife decline and, 40
hybrid receptor "chips," 31
hydrogen cyanide (HCN), 71
hypersensitivity syndrome, 104

I
IEI (Idiopathic Environmental Intolerance),
    99–100
IGF-I (insulin-like growth factor-I), 115
immune response, carcinogenic cells, 31
immune systems, 56
immunology, advances in, 32
immunosupression in aquatic animals, pol-
    luted waters and, 107
immunotoxicity testing, 26
in vitro assays, animal testing and, 26
in vitro data predicting in vivo response,
    31–32
in vitro defined, 31
in vitro studies, chemical effects, 97
inbred laboratory mice, receptors and, 32
inbred laboratory rats, sensitivity levels in,
    14, 16
indole-3-carbinol, cancer and, 64
    broccoli, 66
infant mortality rates, 58
infectious disease, 146
inhalation exposure, Sick Building
    Syndrome, 102
inihibition of acetylcholinesterase enzyme,
    172
injected drugs, 26

insect-borne diseases, DDT and, 42
insecticides, 38
    benefits of, 41
    naturally occurring, 52
    new, biological activity, 92
    toxicology, 174–175
insects
    resistance to pesticides, 52
    threats from, 85
    VIPA example, 85–87
inspections, bioterrorism and, 143
insulin
    bovine, 114
    BST comparison, 113
interactions
    additive, 95
    alcohol and other chems, 30
    antagonistic, 95
    chemical mixtures and, 96
    drinking water, 96
    smoking and alcohol consumption, 96
    synergistic, 95
    tetracycline/calcium, 98
    tobacco and cancer, 29–30
IPM (Integrated Pest Management)
    techniques, 51, 122
Iraq
    chemical warfare, 138
    cyclosarin, 138
    sarin, 138
    sulfur mustard, 138
It Ain't Necessarily So, xiii

K
kanamycin, resistance in plants, 131
kidney toxicity, 13–14
kidneys
    absorption and, 26–27
    Balkan nephropathy, ochratoxin and, 76
    damage in rats
        amnioglycoside drugs, 15
        gentamicin and, 15
kindling, chemical sensitization and, 105
Kligerman, drinking water in rats/mice, 96
Kuru, human flesh and, 146

L
laboratories, natural vs synthetic, 71
laboratory mice, drinking water and, 96

laboratory rats
  drinking water, 96
  inbred status, 14
    sensitivity levels, 16
  selection, 14
Laetrile, amygdalin and, 72
Landenberg, coiine, 70
lead arsenate, 41
legal issues
  circumstances and, 88
  FDA guidelines and, 12
  future exposure, 80
  MCS and, 80
  psychology of exposure, 80
lewisite, chemical warfare, 138
LHRH (luteinizing hormone releasing
    hormone), 114
life-expectancy, Census Bureau information,
    57–58
lifetime human consumption, tolerances
    and, 45
limonene, 74
lindane, 166, 169–170
linuron, 49
lipophilic nature of halogenated hydro-
    carbons, 167
liver
  absorption and, 26–27
  alcohol consumption, 28–29
    chronic, 96
  analgesics and alcohol, 96
  damage results, 61
  enzymes, toxicity and, 27
  increased toxicity of chemicals, 29
  liver disease
    cirrhosis and alcoholism, 59
    decrease, Silent Spring and, 59
low-dose exposure
  accumulation, 97
  allergic reactions and, 32
  benefits, Dr. Edward Calabrase and,
    66–67
  dose-response paradigm and, 32
  hypersensitivity and, 32
  mathematical models for extrapolation, 93
lung cancer, 60
  cancer rate increase and, 60
  fruits & vegetables consumption, 64
  phenethyl isothiocyanate and, 66

lutein, 65
lycopene, 65

M
ma huang, side effects, 71
Mad Cow Disease, 146
malaria, DDT and, 42
malathion, 49
maneb, 49
mango juice, natural toxins, 74
mastitis, BST-treated cows and, 116
mathematical models for extrapolation, 83
MCS (multiple chemical sensitivity)
    syndrome, 32, 40, 99
  allergy similarity, 102
  compulsive behavior, 103
  defining syndrome, 101
  depression and, 103
  food allergies and, 102
  legal issues and, 80
  nonfunctional immune systems, 104
  pets and, 105
  psychological causes, 103
  psychological factors, 103–105
  sensitivity and, 100
  smell, role of, 103
  smell and, 104
  symptoms, 101
  taste and, 103
  triggers, 102
meat-bone meal as supplements, 146
mechanism of action, 32
media
  death from natural causes versus drugs or
    chemicals, 4
  misinterpretations, 22
melatonin, 31
metabolic conversion of halogenated
    hydrocarbons, 167–168
methoxypsoralen, 74
methyl isocyanate, explosion in India, 71
methyl mercury, robins and, 40
methyl-parathion, 49
MFO (mixed-function oxidase) enzyme, 168
mice. See laboratory mice
microbiology, 146
milk
  BST presence, 113
  discarded after medical treatment, 117

milk (*continued*)
  higher production, mastitis and, 116
  increased production, 112
  low fat, pesticides and, 118
  natural botulism outbreak, 142–143
  residues in, 117
  tanker trucks, bioterrorism, 143
millet, 126
mink, transmissible mink encephalopathy,
  146
mirex, 166
  phase out, 42
mixtures, chemical. *See* chemical mixtures
modern medicine, death rate and, 61
  cancer survival rates, 61
mold
  mycotoxins and, 75
    animal feed, 77
    peanuts and, 74
molluscicides, 38
monarch butterfly larvae,
  Bt corn consumption and, 131
Monsanto, 111–112
morning sickness, 17
mortality, 56–57
MRI (maximum residue limit), 164
MTD (maximum tolerated dose), 160–161
mustard gas shells, 138
mutagenicity, Ames test, 73
mycotoxins, 75
  contaminant classification, 75
  Delaney Clause and, 75–76
  estrogen and, 76–77
  foodborne attacks, 142
  natural, 77

N
natural carcinogens, 28
  coffee, 78
  synthetic chemicals and, 41
natural food industry, 69
natural foods' MTD, 161
natural hormones, 74
natural pesticides vs synthetic,
  consumption, 78
natural toxins, 74
nerve agents
  gases, 141
  VX, 143

Nesse and Williams, 98–99
neurotoxins, naturally occurring, 52
New Variant Creutzfeldt-Jakob disease, 146
Newsweek article, feedback, xvi
NIH (National Institutes of Health), animal
  testing and, 15
nitrate, fertilizer, 52–53
no-effect levels, tolerance, 89
NOAEL (no observed adverse effect level),
  160
nonfunctional immune systems, MCS and,
  104
nutraceuticals, 71
nutritional substitutes, regulations and, 71

O
Objective Scientists Model, xiv
occupational exposure to pesticides, 19
  death, acute, 42
  farmers' cancer rates and, 61–62
ochratoxins, 76
  Balkan nephropathy and, 76
off-gasing of formaldehyde from building
  materials, 103
older persons, cancer rates in, 60, 78
oncogenes, 16
oral administration, peptides, 114
orange juice, natural toxins, 74
organic cynaide gasses, 71
organic farming
  fertilizer, urban sewage sludge, 52–53
  product labeling, 53
  technique development, 52
organophosphates, 38, 170–174
  accumulated exposure, 97
  acetylcholinesterase, 97
  acute deaths, 42
  diisopropylfluorophosphonate, 137
oriental medicine, 69
over-the-counter remedies, 95

P
Paracelsus, dose and, 19
paraoxon, 171
Parkinson's disease, rotenone and, 52
parsley, natural toxins, 74
pasteurization, BST and, 115
PCBs (polychlorinated biphenyls), 29, 166
  breast cancer and, 107–108

estrogen resemblance and, 107
  phase out, 42
PCP (pentachlorophenol), 168, 170
peanut hazards, 74
  aflatoxins, 76
peptide hormones, levels detected, 114
peptides, 113
  absorption, 113–114
  digestion and, 113
  oral administration, 114
Peregrine falcon, decline in numbers, 39, 40
persistence, 165–166
  new insecticides, 92
persistent chemicals, 39
pesticide chemists, future developments
    and, 92
pesticide-induced cancer
  Doll and Peto, 61–62
  Scheulplein and, 61
pesticides
  actual usage assessment, 49
  airlines' use, 95
  benefits of, 41
  cancer epidemic and, 105
  definitions, 37
  detection vs pesticide-induced toxicity,
      48
  exposure
    average household, 94
    detection, 103
  food additive classification, 75
  insects' resistance, 52
  modern causes, 44
  nonfood crops, 49
  occupational exposure, 19
  personal repellants, 94
  pets, 95
  placebo effect, 11
  pounds vs grams per acre, 51
  proposed phaseout, 42
  reduction in amounts, 51
  scare, origins of, 38
  sperm count decreases and, 108
  Top 25 for 1974 and 1993, 50
  unregistered, FDA targeting, 46
  vegetarians and, 64–65
pests
  definitions, 37
  resistence in plants, 126

pets
  MCS and, 105
  pesticide toxicity, 95
pharmacokinetics, drug doses and, 15
phenethyl isothiocyanate, lung cancer and,
    66
phenotypes, 123, 124
  plant breeders and, 124
phosgene, chemical warfare, 138
phosphates, 171
phosphonates, 171
phosphonothionates, 171
phosphoramides, 171
phosphorothionates, 171
pigs, zearalenone, 77
placebo effect
  drug trials, 10
  pesticides and, 11
plant biotechnology, 123, 124
  Bt, 126
  Calgene FlavrSavr tomato, 127
  Cry9C enzyme, 126
  European corn borer, 126
  golden rice, 126
  Roundup, 126
  science fiction and, 132
plants as poisons, 70
political polls, 7
population, statistics and, 5
  death rate reporting, 59
Posilac, 112
posphonothionothiolate, 171
potassium cyanide (KCN), 71
pounds per acre, pesticides, 51
premature thelarche, zearalenone and, 77
prescription drugs, 95
Preston, Richard (The Hot Zone), 150
prions, 146
  humans and, 146
produce. See also fruits & vegetables
  Alar, 49
  consumption rates, 46
  FDA Residue Monitoring Prograon, 44
  residues, 46
  violative residues, 45
product labeling, organically grown foods,
    53
propachlor, 49
propanil, 49

propoxur, 171
proteins
    genes inserted in transgenic plants, 129
    prions, 146
proteomics, 125
provability of effects, 93
psoralens, celery and, 52
psychological causes of MCS, 103
psychological factors, MCS and, 103–105
psychology of exposure, legal issues and, 80
psychology of risk perception, 83
PTSD (Posttraumatic Stress Disorder), 100
public health, mycotoxins and, 77
pyrethrins, 52, 174
pyrethroids, 38

Q–R
Q fever, food contamination and, 139

radiation-induced cancer, Delaney Clause
    and, 33
Randolph, Dr. Theron, MCS, 101
rats. *See* laboratory rats
Ray, Dixie Lee, DDT withdrawal, 41
rbGH, 112
rbST, 112
reality hierarchy, 35
receptors
    antioxidants and, 30–31
    binding to, dose-response and, 30
    binding to endogenous compounds, 30
    chemical binding, 31
    computer chips and, 32
    DNA gene "chips," 31
    hybrid, 31
    inbred mice, 32
    receptor assays, in vitro studies, 31
    signal transduction and, 30
    switches description, 31
Red Army Faction, botulinum toxin and,
    141
Red Dye #2, 92
reductionism, risk assessment and, 79
regulations
    aggregate exposure, 94
    bioterrorism and, 143
    new, applied to old events, 88
regulatory agencies
    bans, evidence for, 84

focus, 70–71
nutritional substitutes, 71
requirements, 161
risk assessment and, 4
relationships and causation, 63
relative risk (side effects), spontaneous rate
    and, 16
reproducibility and consensus, 64
research, directions for, 93
residue
    antibiotics, 38
        in milk, 117
    carcinogenic drugs and, 34
    detectable, 45
    detection methodology, Office of
        Technology Assessment and, 48
    EPA tolerances, 45
    MRI, 164
    produce, 46
    residue level risk, 34
    safety, 32
        farmers, 33
        veterinarians and, 33
    trace, 45
    vegetarian diets, 65
    violative, 45
        consumption rate table, 47
Residue Monitoring Program, FDA, 44
restrictiveness, availability and, 26
retin A, 19
retinol, 20
    carotenoids and, 65
reversal of progress, synthetic drugs and, 70
RfD (reference dose), 162
    hazards and, 163
Rheumatic Condition, Boer War, 100
rice, golden rice, 126
ricin, food contamination and, 139
risk
    exposure and access, 87
    toxicology tolerances, 160
risk assessment
    adverse reaction, 12
    human interpretation and, 11
    media misinterpretation, 22
    methods, 78–79
    reductionism, 79
    regulatory agencies and, 4
    value judgements and, 12, 83

risk perception, 83
risk ratio calculation, 16
RNA (ribonucleic acid), 125
  viruses and, 125
rodenticides, 38
rotenone, 52, 174
Roundup, 51, 126

S

saccharin, 92
safety factors, toxicology tolerances and, 160
safrole, 74
salmonella, food poisoning and, 131–132
sarin, 171
  Iraq, 138
Scheuplein, Dr. Robert, pesticide-induced
    cancer, 61, 78
Schrader, Dr. Gerhard, 171
science fiction, plant biotechnology and,
    132
scientific results misinterpretation, 4
scrapie, sheep and, 146
semantics, 99
senescence, aging, 98–99
sensitivity
  analytical chemistry, 33
  chemical sensitivity, 101–102
  MCS and, 100
  residue tolerances, EPA guidelines, 45
  synthetic chemicals, chronic exposure,
    102
sensitivity of approach method,
    residue regulation, 34
sewage sludge, organic fertilizer, 52–53
sexual functioning in wildlife, 107
sheep, scrapie and, 146
Shell-Shock, 100
Sick Building Syndrome, 99–100, 102
  inhalation exposure, 102
  off-gasing of formaldehyde, 103
side effects risks
  anthrax vaccine, 12
  chance occurrences, 13
  diet pills, 71
  ephedra alkaloids, 71
  FDA and, 12
  separating from disease, 13
  tolerance for greater results, 93
  vitamin A predictions, 22

signal transduction, receptors and, 30
Silent Spring, 38–40
  benefits of, 43
skin cancer, 60
skin toxicity, organophosphates, 42
smell
  MCS and, 104
  role of in MCS, 103
  Sick Building Syndrom and, 102–103
smoking
  alcohol consumption and, 96
  cancer and, 60
Society of Toxicology, DDT ban and, 43
sociology of risk perception, 83
Socrates, hemlock, 70
sodium chlorate, 49
solanine, 52
Soldier's Heart, Civil War, 100
soman, 171
sometribove, 112
sorghum, 126
sperm count decreases in past 50 years,
    106–107
  pesticides and, 108
spontaneous cures/diseases, 11, 15
  birth defects, 17
  linking to drug or chemical exposure, 16
Sprague-Dawley rats, 14
Staphylococcal enterotoxin B, food contam-
    ination and, 139
Staphylococcus, food poisoning and,
    131–132
StarLink transgenic corn, 127
  allergies and, 130
statistics
  applications
    medicine, 4
    science, 4
  bell-shaped distribution, 5
  bimodel population and, 5
  chance and, 5, 9
  clusters, gentamicin testing in rats, 15
  coin flip example, 9
  confidence interval, 7
  drug trials, 9–14
  polling example, 7
  population, 5
  principles, 4–8
  variability and, 5

strabismus, Botox and, 141
stress, veterans' illnesses, 100
Strigia, 126
studies' purposes and limitations, 99
subchronic effects, 160–161
subtoxis chemical exposure, 22
sulforaphane, cancer and, 64
    broccoli, 66
sulfur mustard, chemical warfare, 138
    Iraq, 138
super weeds, 131
supplement industry, 69
Sydney funnel-web spider, 85–87
synergistic interactions, 95
    mixtures, 98
synthetic chemicals
    natural carcinogens and, 41
    sensitivity, chronic exposure and, 102
synthetic drugs' natural origins, 70
synthetic fungicides, peanut mold, 74–75
synthetic pesticides vs natural,
        consumption, 78

T
T 1/2 (half-life), 165–166
tabun nerve agent, 138, 171
taco shells, transgenic corn, 127
Tarheel accusation, xvi
taste, MCS and, 103
TCDD, 175
TEPP (tetraethylpyrophosphate), 171
teratogens
    natural, nausea in pregnancy and, 17
    naturally occurring, 52
terrorism. *See* bioterrorism
testicular cancer, increase in, 107
tetrachlorvinphos, 171
tetracycline, 98
The Causes of Cancer, 61–62
*The Death of Common Sense: How Law is
        Suffocating America*, xiv
The Edge of the Sea, 39
The Hot Zone, 150
The Sea Around Us, 39
thiolates, 171
thresholds
    consumption assumptions, 48
    Delany Clause and, 33
    exposure, paradigm shift, 88

high-dose studies, 48
    toxicology tolerances and, 160
    violative levels and, 48
thyrotropin releasing factor (TRF), 114
tissue, absorption and, 27
tobacco, cancer and, 29–30
tolerances
    defined, 45
    EPA, pesticides, 164
    establishment, 160
    exceeded amounts, 45
    no-effect levels, 89
tomatoes, Clostridia botulinum, 141
toxaphene, 49, 166
    phase out, 42
toxicity
    amino acids, 23
    chemical mixtures, 94
    detection and, 34
    false impressions, 105
    increased in liver, 29
    liver enzymes and, 27
    skin toxicity
        organophosphates, 42
        psoralens, celery handling, 52
    transgenic plants, 129
    vitamin A, 20
    vitamins, 23
    VX nerve agent, 143
    water barrel analogy, 97
toxicology
    allergen tests, 131
    chemical mixtures, 99
    future of, 93
    herbicides, 175–178
    improper use, dose-response and, 29
    insecticides, miscellaneous, 174–175
    mycotoxins, 75
    organophosphates, 170–174
    outdated, 44
    tolerances, establishing, 160
    virtual, 81
toxins
    biosecurity and, 135
    natural toxins, 74
    trace residue, EPA tolerances, 45
transgenic plants, 127
    allergens and, 130
    ecological concerns, 131

genetics, 129
  proteins inserted, 129
  toxicity concerns, 129
transgenic technology, 125
transmissible mink encephalopathy, 146
trichothecenes, 77
typhoid fever, DDT in WWII, 41

U
U-shaped dose response phenomenon, 24
unregistered pesticides, 46
USDA (U.S. Dept. of Agriculture)
  produce classifications, 44
  tolerance setting, 164

V
value judgments, risk assessment and, 12
vapor pressure, chlorinated-hydrocarbons,
    167
variability, statistics and, 5
vegetables. *See* fruits & vegetables
vegetarians, cancer rates and, 64–65
vesicants, chemical warfare, 138
veterans of war, illnesses, 100
veterinarians, residue safety and, 33
violative residue, EPA tolerances, 45
  consumption rate table, 47
  drinking water, 62
  produce, 45
VIPA (Venomous Insect Protection
    Administration) example, 85–87
viral encephalitis, 42
virtual toxicology, 81
viruses, transfecting cells and, 125
vital statics, Census Bureau, 57
vitamin A
  absence of, 20

cancer argument, 23
  derivatives, 19
  high exposure, low doses, 20
  predictions of risk, 22
  thresholds, 21
vitamin B, 65
vitamin C, 65
vitamin D, 65
vitamin E, 65
VX nerve agent, 143

W–Z
war. *See* biological warfare; chemical
    warfare
water barrel analogy, 97
waterbourne bacterial epidemics, chlorine
    in water and, 43
West Nile Virus, 42
Western society's idea of control over
    disease, 4
*Why We Get Sick: The New Science of
    Darwinian Medicine*, 98–99
wildlife
  decline, human population and, 40
  sexual functioning, 107
Witchweed, 126
wrinkles, Botox and, 141
WWI, chemical warfare, 137
WWII
  chemical warfare, 138
  DDT use, 41

zearalenone, 76–77
  premature thelarche and, 77
zero tolerance policy, Delaney Clause, 34